BIOCHEMISTRY RESEARCH TRENDS

MEDICAL ENZYMOLOGY: A SIMPLIFIED APPROACH

BIOCHEMISTRY RESEARCH TRENDS

Additional books in this series can be found on Nova's website under the Series tab.

Additional E-books in this series can be found on Nova's website under the E-book tab.

MEDICAL ENZYMOLOGY: A SIMPLIFIED APPROACH

TAREK H. EL-METWALLY
AND
YAKOUT A. EL-SENOSIL

First Edition

Nova Biomedical

Nova Science Publishers, Inc.
New York

For permission to use material from this book please contact us:
Telephone 631-231-7269; Fax 631-231-8175
Web Site: http://www.novapublishers.com

Additional color graphics may be available in the e-book version of this book.

LIBRARY OF CONGRESS CATALOGING-IN-PUBLICATION DATA

El-Metwally, Tarek H.
Medical enzymology : a simplified approach / Tarek H. El-Metwally.
p. ; cm.
Includes bibliographical references and index.
ISBN 978-1-60876-904-9 (hardcover)
1. Clinical enzymology. 2. Enzymes. I. Title.
[DNLM: 1. Enzymes--physiology. 2. Biocatalysis. QU 135 E48m 2010]
RB48.E46 2010
616.07'56--dc22

2009054165

Published by Nova Science Publishers, Inc. ✝ New York

CONTENTS

With love to our families

AUTHORS

Tarek H.El-Metwally,
Professor of Medical Biochemistry, Molecular/Cellular/Cancer Biology and Consultant Clinical Biochemist, Department of Medical Biochemistry, Faculty of Medicine, Assiut University, Assiut, Egypt.

Yakout A. El-Senosi,
Associate professor of Biochemistry and Consultant Clinical Biochemist, Department of Biochemistry, Faculty of Veterinary Medicine, Benha University, Benha, Egypt.

LEARNING OBJECTIVES

After reading this book the student should be able to:

- Describe the characteristics of enzymatic reactions from the viewpoint of free energy, equilibrium, kinetics and direction of the reactions in comparison with simple chemical reactions.
- Discuss the structure and composition of enzymes, including apoenzyme, coenzymes, cofactors, and prosthetic groups, and conditions that affect the rate of enzymatic reactions.
- Describe enzyme kinetics based on the Michaelis-Menten equation and the significance of the Michaelis constant (Km).
- Describe the elements of enzyme structure that explain their substrate specificity and catalytic activity.
- Explain the global regulatory mechanisms affecting enzymatic reactions, including regulation by allosteric effectors and covalent modification.
- Differentiate among the three types of enzyme inhibition from the viewpoint of enzyme kinetics.
- Explain the bases of the therapeutic use of enzyme inhibitors, different methods of measurement of the enzymatic activity and the diagnostic utility of clinical enzyme assays.
- Describe the different approaches of enzyme engineering and design and their applications.

Chapter I

IMPORTANCE OF ENZYMOLOGY

INTRODUCTION: NONENZYMATIC VS. ENZYMATIC CATALYSIS

The dynamic changes in cellular and integrated body functions through constant changes in their chemical composition are largely due to the regulated action of enzymes. Therefore, rate of a specific cocktail of regulated enzymatically catalyzed reactions defines a cell and the living organisms at large.

Reactions occur in biological systems rapidly under very mild pH (~7), temperature (37 oC) and pressure due to catalysis. Such catalysis in carried out by enzymes as biomolecular organic biological catalysts produced by and found in living organisms (including some viruses) that enhance rate of chemical reactions. However, in optimized in vitro conditions, they also work independent of the cells that produce them. Among the two fundamental prerequisites of any form of life is efficient and specialized catalysis of chemical reactions along with the ability to self-replication that in itself is dependent on efficient and specialized catalysis. Every catabolic or anabolic reaction in the body is catalyzed by an enzyme that is expressed by specific gene(s). About 3000 enzymes are known.

At constant pressure, the uncatalyzed reaction may occur spontaneously (when it is exergonic, i.e., energy-releasing because the products have a lower energy content than reactants) as the case with sodium ionization ($Na \Rightarrow Na+$), occurs at a very slow rate as the case with decomposition of H2O2 into H2O and O2, or, will never occur, as the case with glucose phosphorylation into glucose-6-phosphate on the expense of ATP (when it is endergonic, i.e., energy-requiring because products have higher energy content than reactants). The rate of the catalyzed phosphorylation of glucose (3 mM) into glucose-6-phosphate utilizing ATP (2 mM) by hexokinase (0.1 μM) is 10-3 μM/sec, whereas, the rate of the non-enzymatic reaction in same conditions is 10-13 μM/sec; i.e., the enzyme made the rate 1010 times faster. This is largely dependent on the thermodynamic nature of the reactants.

Catalysis refers to the acceleration of the rate of a chemical reaction by a substance, called a catalyst. Catalyst itself is not consumed in the reaction but may acquire a reversible change from which it is recoverable. Catalysis is crucial for any known form of life, as it makes a thermodynamically favorable and unfavorable chemical reactions to proceed into biologically relevant much faster rate; sometimes by a factor of several million times. Catalysts accelerate the chemical reaction by providing a lower energy pathway between the reactants and the products. This usually involves the formation of an intermediate, which cannot be formed without the catalyst. The formation of this intermediate and subsequent reaction generally has a much lower activation energy barrier than is required for the uncatalyzed direct reaction of reactants into products. As all catalysts, enzymes do not alter the position of the chemical equilibrium of the reaction. Usually, in the presence of an enzyme, the reaction runs in the same direction as it would without the enzyme, but just more quickly. The enzyme catalyzes the forward and backward reactions equally depending on the concentration of its reactants. Another distinction for enzymes as catalysts is that they couple two or more reactions, so that a thermodynamically favorable reaction drives a thermodynamically unfavorable one. A common example is enzymes which use the dephosphorylation of ATP to drive some otherwise unrelated chemical reaction. On the other hand, chemically catalyzed reactions, e.g., by copper, acids or bases, have several differences with the organic biological catalysts, i.e., enzymes (Table 1).

Although were noticed earlier to be contained in yeast upon studying sugar fermentation by Louis Pasteur, enzymes were named so by F. W. Kühne (1878) for the catalytically active substances existing in the yeast (Greek, en = in, zyme = yeast or ferment). The word enzyme was used later to refer to catalytic molecules extracted from living cells, e.g., pepsin, and the word ferment was used to refer to chemical activity produced by living organisms, e.g., brewer's yeast. Enzymes are thermolabile organic colloidal catalysts of a globular protein nature produced by the living cells for the function of specific catalysis of chemical reactions of specific nature on specific reactants (substrates). Nevertheless, some enzymes are RNA in nature, i.e., ribozymes.

Table 1. Differences between enzymes as biological catalysts and other nonenzymatic catalysts.

	Enzymes	*Chemical catalysts*
1	Thermolabile.	Thermostable.
2	Organic, biological substances.	Mostly inorganic, non-biological substances.
3	Protein in nature, denaturable.	Non-protein, non-denaturable.
4	Different grades of specificity for substrate and nature of the chemical reaction.	Non-specific.
5	Body temperature, pH and pressure are their optimum.	Require unphysiologically high temperature, pressure or extreme pH.
6	High catalytic efficiency by forming enzyme-substrate complex (reaction rate is 10^5-10^{17} greater than uncatalyzed and several orders of magnitude greater than the chemically catalyzed reaction).	Low catalytic efficiency because of absence of real catalyst-substrate complex.
7	Could directly couple a thermodynamically favorable reaction to drive a thermodynamically unfavorable one.	It does not.
8	They are mostly susceptibility to regulation at their gene level and/or the existing molecule level.	Unregulated.

Enzymes are highly specific in their action and act in different compartments inside the cells (i.e., metabolic enzymes) and in the extracellular body fluids and lumens (e.g., blood clotting factors and the digestive enzymes, respectively). They remain chemically essentially unchanged during the reaction, but they speed the rate of advancement towards the equilibrium of the reaction without changing such equilibrium. For example in a reaction A⇔B with a forward rate of 10-4/second and a backward rate of 10-6/second at equilibrium, the reaction equilibrium is kforward/kbackward, i.e., 10-4/10-6

= 100. This means that at equilibrium of the reaction the concentration of B will be 100 times that of A whether the reaction is catalyzed or not. However, if the uncatalyzed reaction would take □1 hour to attain equilibrium, the enzyme catalyzed reaction would require □1 second. Almost all chemical reactions occurring in the body need catalysis by certain enzyme(s) to proceed at significant rates; a very few reactions occur spontaneously after a necessary enzyme activated step. Although enzymes may be involved in the intermediary reactions that transform a substrate into product, they are regenerated to their original pre-reaction forms. The set of enzymes in a cell determines which metabolic pathways occur in that cell (metabolomics).

Extra Attention: Diseases and Enzymes

Mutation in the enzyme-expressing gene(s), defective transcription, post-transcriptional- or post-translational processing, or targeting of the enzyme leads to deficiency of the enzyme activity at the target site, and, thence deficiency of a metabolic reaction product and accumulation of its substrate and/or alternative products. This is the base of the metabolic inborn errors of metabolism as genetic diseases. To date □1400 such defects scattered throughout metabolism were recorded. Sometimes enzyme hyperactivity could be the error, e.g., cancer cell proliferation-related enzymes. Measurement of the enzyme activity in blood plasma, blood cells or tissue samples is important in characterizing these diseases - Clinical Enzymology - and assessment of therapy.

The catalytic activity of an enzyme is not only important for productive intermediary metabolism, recycling and digestion but is also important for cellular activities such as signal transduction and cell regulation. Enzymes are energy-transducing machineries, e.g., photosynthesis transforms the light energy into chemical-bond energy through an ion gradient. Oppositely, mitochondrial oxidative phosphorylation transforms chemical-bond energy of the food component into the free energy to create ion gradient. The gradient is used to derive membrane transport activity and energy recapturing as ATP. Kinases/phosphatases and acetylases/deacetylases regulate cell signalling and DNA activities. ATPases in the cell membrane are ion pumps involved in active transport through creating chemical and electrical gradients. The chemical-bond energy of ATP is transformed into mechanical energy of contracting muscles through, e.g., myosin (heavy chain - a protein serine/threonine kinase; 2.7.11.7, and, light chain – a calcium/calmodulin-dependent kinase; 2.7.11.18). More exotic functions include generating of light such as luciferase (EC 1.13.12.5) in fireflies from chemical-bond energy. Enzymes also enable viruses to infect cells, such as the HIV integrase and reverse transcriptase; or for viral release from cells, like the influenza virus neuraminidase.

Chapter II

SYSTEMS OF ENZYME NOMENCLATURE AND CLASSIFICATION

ENZYME NOMENCLATURE

With the progressive development in understanding the nature and mechanisms of enzymes 4 systems of their nomenclature were adopted.

1. ***Empirical and broad sense naming:*** Initially, the enzymes acquired arbitrary names at the time of their discovery without following any rational rule, e.g., pepsin (EC 3.4.23.1; for Greek pepsis = digestion), trypsin (EC 3.4.21.4; for Greek tryein = tearing pancreas as its source), papain from the papaya fruit and ptyalin. *Lysozyme* is a lysosomal glycosidase that was named so because it is a natural antibacterial agent found in tears, saliva and egg whites. It cleaves the β1-4 glycosidic bond between the two types of sugar residues in the bacterial cell walls peptidoglycan (*N*-acetylmuramic acid and *N*-acetylglucosamine). In this system, it is clear that the name does not specify substrate, product or nature of the reaction. With such ambiguity, it happened that the same enzyme is known by more than one name.
2. ***Substrate-dependent naming:*** The name is derived from the substrate with "-ase" as an enzyme meaning suffix, e.g., urease (for urea), lipase (for lipids) and protease (for protein) after Eduard Buchner (1897). Even though these were meaningful by at least specifying substrates, still some of these do not even do so, e.g., catalase (breaks $2H_2O_2 \Rightarrow 2H_2O$ and O_2).

3. ***Chemical nature of the reaction-dependent naming:*** The name depends on nature of chemical reaction with -ase as a suffix, e.g., hydrolase (for hydrolysis), dehydrogenase (for oxidation by removal of hydrogen), and transaminase (for reversible transfer of amino groups from one amino acid to an α-keto acid).
4. ***Systematic naming:*** Because of discrepancies of these systems and the ever-increasing number of newly discovered enzymes, The Enzyme Commission of the International Union of Biochemistry and Molecular Biology in 1961 constructed a systematic mechanism of action-based classification. Enzymes were subdivided into 6 classes, subclasses and sub-subclasses. Every enzyme is given a written systematic name within each sub-subclass and a code digital identification name. The written name of an enzyme is formed of the substrate/product name, the coenzyme name and the class of the chemical reaction suffixed with "-ase". The digital name of an enzyme is composed of four parts separated by a colon or a full stop (EC W.X.Y.Z), where, EC refers to Enzyme Commission numbering system; W refers to the enzyme class, i.e., the type of reaction catalyzed; X refers to the subclass, i.e., the general substrate or chemical group involved; Y refers to the sub-subclass, i.e., the specific substrate or coenzyme; and, Z refers to the serial number of the individual enzyme among the list of the subsubclass. Example is alcohol:NAD:Oxidoreductase (EC 1:1:1:1) that is; an oxidoreductase (class 1) catalyzes ethanol -OH group oxidation into acetaldehyde (subclass 1 acting on -OH) on the expense of NAD as electron acceptor (subsubclass 1) activated by the enzyme alcohol dehydrogenase *per se* listed number 1 in the subsubclass (1). Another example, EC 2:7:3:2 is the systematic name for ATP:Creatine Phosphotransferase. However, the shorter and more familiar and convenient trivial names of the 1, 2 and 3 naming systems are still widely in use.

$$\underset{\textit{Ethanol}}{CH_3-CH_2-OH} + NAD^+ \xrightleftharpoons{\textit{Alcohol dehydrogenase}} NADH.H^+ + \underset{\textit{Acetaldehyde}}{CH_3-CHO}$$

Classification of Enzymes

According to the systematic mechanism of action-based classification and nomenclature enzymes are divided into 6 enzyme classes as follows:

1. ***EC1 Oxidoreductases:*** They catalyze simultaneously a pair of oxidation and reduction reactions of substrates, where one compound is oxidized and the other is reduced by transfer of protons and/or electrons, e.g., dehydrogenases/reductases, transhydrogenases, oxidases, oxygenases (mono- or di-), and, peroxidases/catalase. *Dehydrogenases/ reductase* catalyzes the reversible/irreversible transfer of hydrogen atoms from donor to acceptor. *Transhydrogenase* catalyzes the reversible transfer of hydrogen between two carriers, e.g., NAD/NADPH transhydrogenase. *Oxidase* removes reducing equivalents from a substrate and use O_2 as acceptor to release H_2O_2 or superoxide as a byproduct, e.g., xanthine oxidase that catalyzes conversion of xanthine + O_2 + H_2O into uric Acid + H_2 O_2. *Oxygenase* incorporates one (*monooxygenase*) or the two atoms (*dioxygenase*) of molecular oxygen into a substrate to produce hydroxide or hydroperoxide, e.g., conversion of arachidonic acid into 5-hydroperoxy arachidonic acid catalyzed by 5-lipooxygenase. The monooxygenase utilizes NADPH or tetrahydrobiopterin to reduce the remaining O_2 atom into water, e.g., phenylalanine conversion into tyrosine catalyzed by phenylalanine hydroxylase. *Peroxidase* reduces peroxides (inorganic H_2O_2 or organic, e.g., lipid peroxides) on the expense of oxidizing a substrate, e.g., glutathione peroxidase (GSH + H_2O_2 $\Rightarrow$ GSSG + 2 H_2O) and paraoxonase. *Catalase* catalyzes oxidation and reduction of H_2O_2 into O_2 or H_2O.

   Systematic examples included in this class include;

   - EC 1.1 (act on the CH-OH group of donors).
   - EC 1.2 (act on the aldehyde or oxo group of donors).
   - EC 1.3 (act on the CH-CH group of donors).
   - EC 1.4 (act on the CH-NH_2 group of donors).
   - EC 1.5 (act on CH-NH group of donors).
   - EC 1.6 (act on NADH or NADPH).
   - EC 1.7 (act on other nitrogenous compounds as donors).
   - EC 1.8 (act on a sulfur group of donors).
   - EC 1.9 (act on a heme group of donors).
   - EC 1.10 (act on diphenols and related substances as donors).
   - EC 1.11 (act on peroxide as an acceptor -- peroxidases).
   - EC 1.12 (act on hydrogen as a donor).
   - EC 1.13 (act on single donors with incorporation of molecular oxygen).

- EC 1.14 (act on paired donors with incorporation of molecular oxygen).
- EC 1.15 (act on superoxide radicals as acceptors).
- EC 1.16 (oxidize metal ions).
- EC 1.17 (act on CH or CH_2 groups).
- EC 1.18 (act on iron-sulfur proteins as donors).
- EC 1.19 (act on reduced flavodoxin as donor).
- EC 1.20 (act on phosphorus or arsenic as donors).
- EC 1.21 (act on X-H and Y-H to form an X-Y bond).
- EC 1.97 (other oxidoreductases).

2. ***EC2 Transferases:*** They catalyze the transfer of a functional group (e.g., phosphate, amino and methyl) or a chemical moiety (e.g., ketole and aldole) from one compound (donor) to the other (acceptor), e.g., aminotransferases, glycosyltransferases, methyltransferases, acyltransferases, phosphotransferases (kinases), transaldolases, transketolases, sulfotransferases and mutases (catalyzes intramolecular transfer of a phosphate group).

Systematic examples included in this class include;

- EC 2.1 (transfer one-carbon groups, Methylase).
- EC 2.2 (transfer aldehyde or ketone groups).
- EC 2.3 (acyltransferases).
- EC 2.4 (glycosyltransferases).
- EC 2.5 (transfer alkyl or aryl groups, other than methyl groups).
- EC 2.6 (transfer nitrogenous groups).
- EC 2.7 (transfer phosphorus-containing groups).
- EC 2.8 (transfer sulfur-containing groups).
- EC 2.9 (transfer selenium-containing groups).

3. ***EC3 Hydrolases:*** They catalyze hydrolytic cleavage of substrates, i.e., breakdown of the compound by addition of water, e.g., thiolases, amidases, ribonucleases, deoxyribonucleases, hydrolytic deaminases, phospholipases, phosphatases, glycosidases, esterase and peptidases. The group removed can be indicated, e.g., adenosine aminohydrolase (EC 3.5.4.4) or the group is suffixed by "-ase", e.g., alkaline phosphatase (E.C. 3.1.3.1). Other than the digestive hydrolytic enzymes, ***lysosomes*** are the major intracellular hydrolytic compartment by hosting hydrolytic enzymes for all complex macromolecules including carbohydrates, protein, RNA, DNA and lipids. These lysosomal enzymes act at optimal acidic pH to recycle molecules. Deficiency of lysosomal hydrolytic enzyme(s) leads to a serious disease(s) characterized by abnormal accumulation of undigested recycling molecule(s) called storage diseases, e.g., carbohydrate and lipid storage diseases.

Systematic examples included in this class include;

- EC 3.1 (act on ester bonds).
- EC 3.2 (act on sugars - glycosylases).
- EC 3.3 (act on ether bonds).
- EC 3.4 (act on peptide bonds - Peptidase).
- EC 3.5 (act on carbon-nitrogen bonds, other than peptide bonds).
- EC 3.6 (act on acid anhydrides).
- EC 3.7 (act on carbon-carbon bonds).
- EC 3.8 (act on halide bonds).
- EC 3.9 (act on phosphorus-nitrogen bonds).
- EC 3.10 (act on sulfur-nitrogen bonds).
- EC 3.11 (act on carbon-phosphorus bonds).
- EC 3.12 (act on sulfur-sulfur bonds).
- EC 3.13 (act on carbon-sulfur bonds).

4. ***EC4 Lyases:*** They catalyze breakage and reformation of bonds in substrates by mechanisms other than hydrolysis or oxidation. Phosphorylases split substrate by adding phosphate, e.g., glycogen phosphorylase, $(Glycogen)_n + H_3PO_4 \Rightarrow (Glycogen)_{n-1}$ + Glucose-1-phoaphate. They include desulfhydrases and dehydratases which reversibly remove/add H_2S or H_2O from substrates, e.g., fumarase (fumaric acid + $H_2O \Leftrightarrow$ malic acid). Non-oxidative decarboxylases: which remove or add CO_2, e.g., pyruvic decarboxylase and other splitters without additions or loses, e.g., aldolases, lyases or cleavage enzymes are lyases. They also convert double bonds into single bonds by adding groups into it, e.g., the aforementioned fumarase reaction.

Systematic examples included in this class include;

- EC 4.1 (carbon-carbon lyases).
- EC 4.2 (carbon-oxygen lyases).
- EC 4.3 (carbon-nitrogen lyases).
- EC 4.4 (carbon-sulfur lyases).
- EC 4.5 (carbon-halide lyases).
- EC 4.6 (phosphorus-oxygen lyases).

5. ***EC5 Isomerases:*** They reversibly catalyze interconversion of different types of isomers that include; *Isomerases:* Cis-trans-isomerase, e.g., trans-Retinol $\Leftrightarrow$ cis-Retinol; and, Aldo-keto-isomerase, Glucose-6- phosphate

⇔ Fructose-6-phosphate. *Epimerases* catalyze interconversion of epimers, e.g., UDP-Glucose ⇔ UDP-Galactose. *Mutases* catalyze interconversion of substrate forms by transfer of a group within a molecule, e.g., glucose-6-phosphate ⇔ glucose-1-phosphate. They are classified also as transferases. *Racemases* catalyze interconversion of D- and L-forms, e.g., D- ⇔ L-Methylmalonyl-CoA.

Systematic examples included in this class include;

- EC 5.1 (racemases and epimerases).
- EC 5.2 (cis-trans-isomerases).
- EC 5.3 (intramolecular oxidoreductases).
- EC 5.4 (intramolecular transferases -- mutases).
- EC 5.5 (intramolecular lyases).
- EC 5.99 (other isomerases).

6. ***EC6 Ligases or synthetases:*** They catalyze covalent C-C, C-O, C-S or C-N new bond formation to ligate 2 molecules together in the presence of ATP (synthetase), e.g., Fatty acid + CoASH + ATP ⇒ Acyl-CoA + AMP + PPi; or in absence of ATP (synthase), e.g., UDP-Glucose + $(Glycogen)_n$ ⇒ $(Glycogen)_{n+1}$ + UDP. Carboxylases are ligases. Some references classify synthases as lyases because it is freely reversible except after PPi (pyrophosphate) hydrolysis into 2Pi (inorganic phosphate) by pyrophosphatase.

Systematic examples included in this class include;

- EC 6.1 (form carbon-oxygen bonds).
- EC 6.2 (form carbon-sulfur bonds).
- EC 6.3 (form carbon-nitrogen bonds).
- EC 6.4 (form carbon-carbon bonds).
- EC 6.5 (form phosphoric ester bonds).
- EC 6.6 (form nitrogen-metal bonds).

Chapter III

Characteristics and Properties of Enzymes

Introduction

The characteristics and properties of enzymes title covers; i) the structural components of the enzyme system; ii) structural-functional classification of enzymes, iii) substrate specificity of enzymes; iv) turnover rate of enzymes; v) enzyme catalysis and mechanisms contributing into it; and, vi) models of enzyme-substrate interaction.

Structural Components and Description of the Enzyme System

Enzymes have the same properties as proteins, e.g., denaturation, precipitation, electrophoresis, etc. The structural integrity of enzymes is a must for their function, therefore, the three dimensional structure-function relationship holds for enzymes like any other protein. Mutations affecting the gene for a specific enzyme may result in disrupted structure-function relationship by one degree or another and consequently results in inherited diseases of different severities.

Structurally, enzymes may belong to *simple proteins*, i.e., formed only of polypeptide chain(s); examples include most carbohydrate and protein digestive enzymes, e.g., pancreatic ribonuclease and amylase. However, most of the known enzymes are classified as metallo-, chromo-, lipo- and glyco-proteins, i.e., *conjugated proteins*. Functionally, enzymes may require the help of certain non-protein factors that may be loosely non-covalently bound (i.e., a coenzyme or cofactor) or firmly covalently or non-covalently attached (i.e., a prosthetic group) to the protein part of the enzyme (i.e., apoenzyme). Enzymes containing a prosthetic group are thus conjugated proteins. Enzymes are generally globular proteins and range from less than 100 amino acid residues in size for a monomer to over 2,500 residues as for the animal fatty acid synthase.

FUNCTIONAL (OR ACTIVE) SITES OF ENZYMES

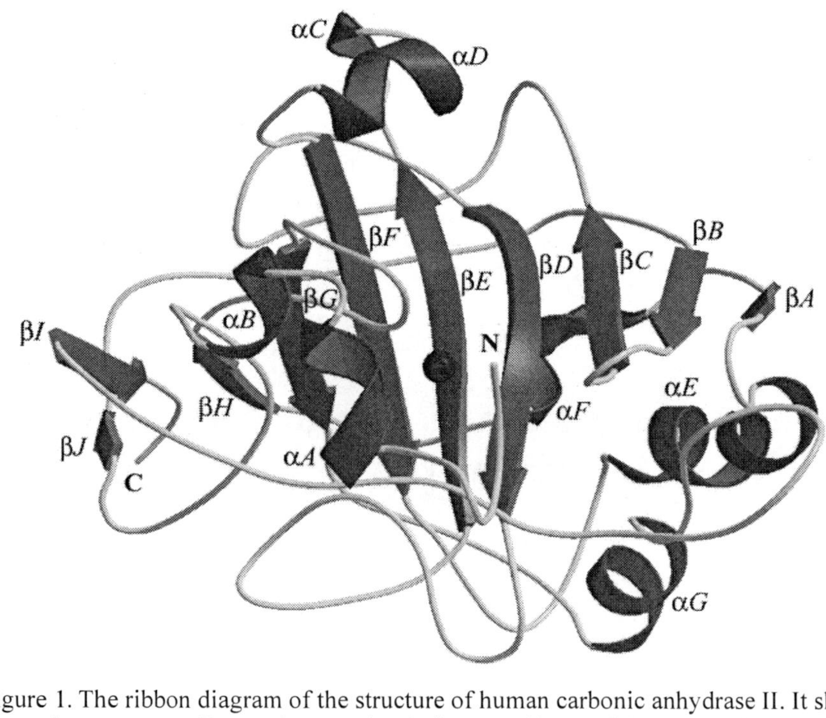

Figure 1. The ribbon diagram of the structure of human carbonic anhydrase II. It shows the tertiary structure (β-strands are red, α-helices are blue, coil is gray, and, the zinc ion at the active site is shown as the central black sphere).

The active sites include; a) The catalytic site; b) The substrate binding-site, and, c) The allosteric site. *The catalytic site* is the region of the enzyme that catalyzes the chemical reaction, i.e., the site(s) which manipulates the substrate to help reaching the reaction transition state and equilibrium faster. It may be slightly separated from the substrate-binding site or they may be integrated into one site. The *substrate-binding site* is the site at which substrate specifically binds and activates the chemical action - along with the catalytic site. The *allosteric site* is an additional binding site that does not have a catalytic function but has a regulatory function on the enzyme substrate binding and/or catalytic functions. The term allosteric site means "the other steering site", i.e., other than and separated from the catalytic/substrate-binding site(s); and, allostery means "a change in shape", i.e., acting like the steering wheel for the car. The non-covalent binding of an allosteric effector at the allosteric site causes a conformational change in the enzyme particularity at the active site(s) that decreases or increases the enzyme activity. Allosteric effectors are substances of low molecular weight with or without structural similarity to substrate. Allosteric effector is called *negative allosteric effector* (or *feedback inhibitor*) when the resulting conformational change decreases the enzyme activity. However, the allosteric effector is called *positive allosteric effector* (or *feedback activator*) if the resulting conformational change increases the enzyme activity (Figure 2, and, allosteric kinetics later).

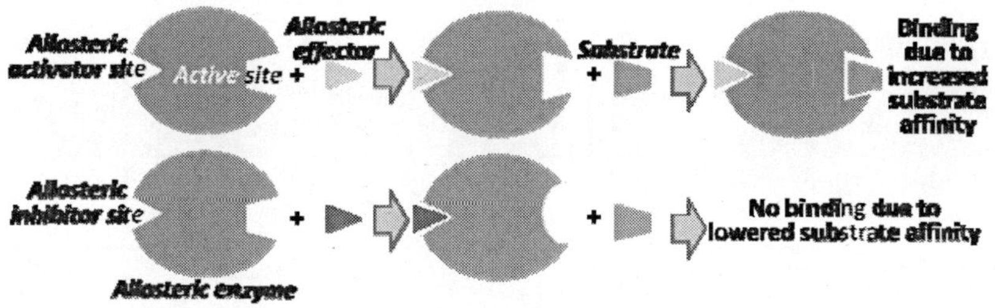

Figure 2. A simplified model of the differential effect of an allosteric activator (upper panel), and, an allosteric inhibitors (lower panel) on the enzyme catalyzed reaction.

The Holoenzyme:

Most enzymes in their *apoenzyme* form require certain obligatory components to help achieving their action and in their absence the reaction will not proceed. These helpers include; *coenzymes, cofactors* and *prosthetic groups* that also participate in the chemical catalysis without being consumed in the reaction. All of them are non-protein in nature, dialyzable (if freed from the apoenzyme) and relatively thermostable organic or inorganic compounds. They function as carriers for reaction substrate(s), reaction intermediates and/or reaction products. Differences between holoenzyme components are presented in Table 2.

Table 2. Differences between the different components of holoenzyme system.

	Apoenzyme	Coenzyme	Prosthetic Group	Cofactor
Nature	Protein	Organic, non-protein	Organic and inorganic	Inorganic
Source	Specific gene	Vitamins or nucleotides	Vitamins, heme and inorganic elements	Inorganic elements
Examples	All enzymes	NAD, FAD, TPP, ATP, UTP	FMN, iron-Heme	Mg^{2+}, Ca^{2+}, Cu^{2+}, Mn^{2+}
Attachment to the apoenzyme		Loose (non-covalent)	Very tight (covalent or non-covalent)	Loose
Heat stability	Labile	Fairly stable	Stable	Very stable
MW	Largest	Smaller	Smaller	Smallest
Determine specificity	Yes	No	No	No
Determine chemical nature of the reaction	Yes	Yes	Yes	Yes

The coenzyme is an organic compound - mostly vitamin-derived or a free nucleotide, e.g., ATP, cAMP, UTP. Examples of vitamin coenzymes are CoASH, TPP, NAD, and FAD. The coenzyme is loosely (i.e., non-covalently) attached to the apoenzyme (ionic and hydrogen bonds and hydrophobic interactions) and is not consumed in the reaction. Since coenzymes are chemically changed during the enzyme action, it is rational to consider them as a special class of substrates, or second substrates. However, coenzymes are usually regenerated into the pre-reaction state and their concentration is maintained at a steady level inside the cell. Examples of coenzymes (See also, vitamins) include: *Hydrogen carriers:* e.g., NAD^+ and $NADP^+$ from nicotinic acid, FAD^+ and FMN^+ from riboflavin, lipoic acid, coenzyme Q, vitamin C and glutathione. *Carriers of moieties other than hydrogen:* include; coenzyme A-SH (CoASH) is an acyl (acid) carrier from pantothenic acid, thiamin pyrophosphate (TPP) is a CO_2 and ketole moiety carrier from thiamin, biotin is a CO_2 carrier, pyridoxal phosphate is an amino group ($-NH_2$) carrier, folic acid is one carbon group carrier, cobalamin is a methyl group carrier. *Energy and phosphate donors:* include; ATP, GTP, UTP, CTP. *Glutathione* is a tripeptide hydrogen carrier and participates in other very important functions *(See, protein chemistry and the pentose monophosphate pathway). Coenzyme Q (Ubiquinone, CoQ)* is a hydrogen carrier utilized in oxidative phosphorylation and is related to vitamin K in the structure. It was named so because it is ubiquitously expressed in large amounts in cells of different tissues particularly the inner mitochondrial membrane.

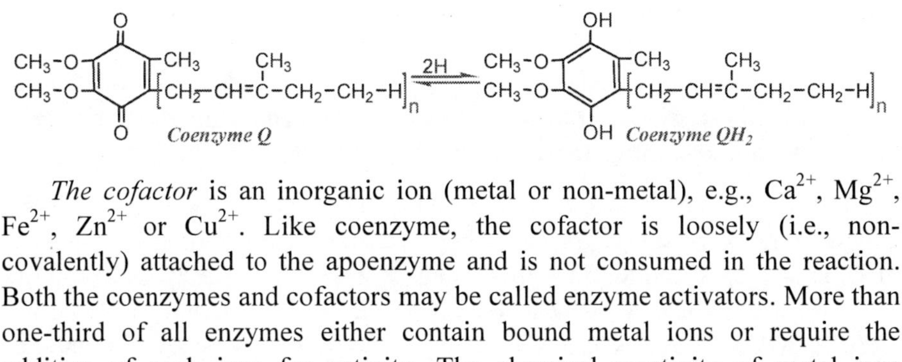

The cofactor is an inorganic ion (metal or non-metal), e.g., Ca^{2+}, Mg^{2+}, Fe^{2+}, Zn^{2+} or Cu^{2+}. Like coenzyme, the cofactor is loosely (i.e., non-covalently) attached to the apoenzyme and is not consumed in the reaction. Both the coenzymes and cofactors may be called enzyme activators. More than one-third of all enzymes either contain bound metal ions or require the addition of such ions for activity. The chemical reactivity of metal ions associated with; their positive charges, with their ability to form relatively strong yet kinetically labile bonds, and, in some cases, with their capacity to be stable in more than one oxidation state - explains why metal ions catalytic strategies were adopted in biological systems. *Therefore, what do metal ions do for the enzyme catalysis?*

- Metal ions could stabilize the 3-dimentional structure of the enzymes and hence contribute to the final conformation required for the reaction.

- Metal ions are mostly found at the active site and hence help in interaction with the substrate.
- Some metals found in the enzymes are transition state elements and thus have multiple oxidation states. Therefore, they can accept or donate electrons during the reaction.
- They could form ternary complexes with the enzyme or substrate.

The prosthetic group is an inorganic element or a complex organic compound or both. The main distinction of prosthetic group from coenzyme and cofactor is being firmly and permanently bound (by covalent or non-covalent strong interaction) to the apoenzyme during the final folding of its protein and its removal irreversibly inactivates the enzyme. They include; selenium in glutathione peroxidase, heme group (is iron-protoporphyrin) in catalase and cytochrome-dependent enzymes, and vitamin-derived coenzymes (e.g., FAD and FMN) in flavoenzymes. Elements such as Zn^{2+}, Cu^{2+}, or Mn^{2+} are prosthetic groups in different isoenzyme forms of superoxide dismutase. Depending upon the attachment of the metal ions with the apo-enzyme the enzyme could be grouped into:

- Metal-Activated Enzymes: The metal ions are associated but not bound to the enzyme and hence can be removed without causing any denaturation or change in 3D structure of the enzyme.
- Metallo-enzymes: Metal is tightly bound and cannot be removed without disrupting the apo-enzyme architecture.

The coenzymes, cofactors and/or prosthetic groups carry, removed/bring groups from/for the substrates and/or products, e.g., removal of hydrogen from succinic acid by succinic acid dehydrogenase to produce fumaric acid and the 2H is released bound to FAD to form FADH2. If FAD is not available the reaction will not proceed. FAD is regenerated by giving 2H to another acceptor, e.g., Coenzyme Q (CoQ + FADH2 $\Rightarrow$ $CoQH_2$ + FAD). Therefore, the coenzyme, cofactor or prosthetic group participates in the catalysis process without being consumed in the reaction.

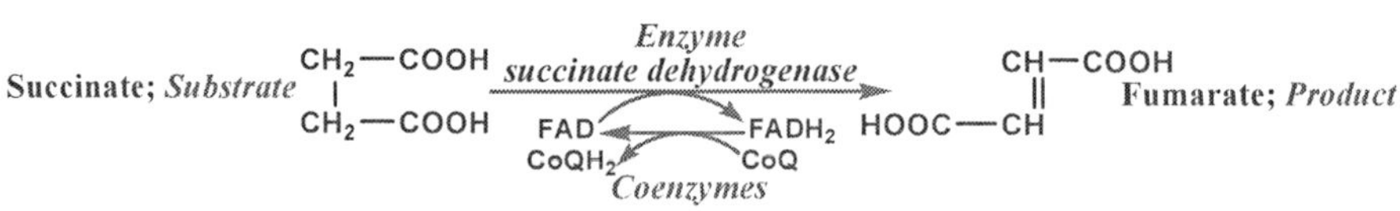

The apoenzyme (apoprotein) is the protein part of the enzyme that determines the substrate specificity and the major determinant of the nature of the chemical reaction. An *enzyme system* or *the holoenzyme* is the form of the enzyme that could accomplish catalysis of a reaction and may be formed from the simple protein apoenzyme only, e.g., most digestive enzymes, or formed of the apoenzyme attached to the coenzyme, cofactor and/or prosthetic group, e.g., most metabolic enzymes. The term "holoenzyme" can also be applied to enzymes that contain multiple protein subunits, such as the DNA polymerases; here the holoenzyme is the complete complex containing all the subunits needed for activity.

EXTRA ATTENTION: RIBOZYMES

Ribozymes, although the overwhelming majority of enzymes are protein in nature, some enzymes are RNA in nature. They are very essential in hydrolysis of phosphodiester bonds of RNA to remove non-coding sequences (introns) as a part of other processes maturating the mRNA. During protein synthesis in ribosomes, the peptidyl transferase activity is thought to be the 28S ribozyme of 60S subunit. Designed ribozymes are used in gene therapy to hydrolyze specific mRNA to prevent its protein expression. Thus, ribozymes have both protein and RNA as substrates and their reaction kinetics with and without inhibitors applies the general kinetics for protein enzymes.

STRUCTURAL-FUNCTIONAL CLASSIFICATION OF ENZYMES

Considering the structural-functional relationship of an enzyme, enzymes are either Key regulatory or non-key regulatory; single reaction enzyme, multifunctional enzyme or multienzyme complex. Key regulatory enzyme is a highly controlled enzyme at levels of its gene expression and its available molecules. It regulates (governs) the rate and direction of the flux of a synthetic or catabolic pathway against other opposing and competing pathways. Their reaction could be a pathway in itself. Their irreversible reaction requires another key regulatory enzyme to reverse its direction. The key regulatory enzymes include a narrower type that is called *the rate-limiting enzymes* or "*pace maker*" that is a key regulatory enzyme with the lowest activity among one pathway sequence of metabolic steps. The rate-limiting step within a cascade of reactions in a pathway is the step requiring highest activation energy.

The key regulatory enzymes usually have relatively low activity, catalyze irreversible metabolic reactions, and frequently are the first or the last in reaction sequences. Key regulatory enzymes are frequently the targets for multiple regulations including; feed-back/exhibit allosteric properties, nutritional and hormonal regulations to insure maximum economy; or are interconvertable enzymes (by covalent modifications) and may have an isozyme pattern. Their gene mutations are the most implicated in inherited metabolic inborn errors. Examples of key regulatory enzymes include; hexokinase, phosphofructokinase, glucose-6-phosphate dehydrogenase, glycogen synthase, hormone-sensitive triacylglycerol lipase and hydroxy-methyl-glutaryl-CoA reductase. Therefore, the biological role, place within the pathway, and regulatory properties are the main criteria that distinguish a key regulatory enzyme.

On the other hand, enzymes that catalyze reversible equilibrium reactions; are shared by more than one pathway (synthetic and catabolic); exhibit high activities; and are present in excess, are called *non-key regulatory enzymes*. Their activities are usually not markedly affected by nutritional, hormonal or neoplastic transformation of the cell. Examples of non-key regulatory enzymes include; phosphohexose isomerase (EC 5.3.1.9) and lactate dehydrogenase (EC 1.1.1.27).

The multifunctional enzymes are monomeric proteins formed of one subunit (i.e., polypeptide) with different domains (or subdomains) each has a separate and different enzyme activities. Within the enzyme intermediary substrate forms are directly transferred without appearing in a free form. Domains are defined regions or subregion of the molecule with defined conformation separated from each other by random coils. Examples are the single subunit of animal fatty acid synthase (with all the required 6 enzymatic activities and one acyl carrier protein function; See Figure 2); eukaryotic acetyl-CoA carboxylase (with 2 enzymatic actions; carboxylation of biotin involving ATP, and, transfer of the carboxyl to acetyl-CoA), DNA polymerase I (with 3 enzymatic actions; polymerase, and, 3'-5'- and 5'-3'-exonuclease), and phosphofructokinase-2 (with 2 enzymatic actions; kinase and phosphatase). Because of their mutlifunctionality, the classification of each enzyme within this group belongs to several classes, e.g., the animal fatty acid synthase catalyzes reactions of EC 2.3.1.38, EC 2.3.1.39, EC 2.3.1.41, EC 1.1.1.100, EC 4.2.1.61, EC 1.3.1.10 and EC 3.1.2.14.

The multienzyme complexes are stable multimeric proteins formed of two or more units (polypetides) with different enzyme activities that are tightly bound and coordinated in one complex by non-covalent interactions. Intramolecularly, they deliver the intermediate substrate products of one activity to the other without such intermediary compounds being free to establish equilibrium in a medium. This allows the generation of very high local concentration of an intermediate thereby enhancing rate of reactions. This organization makes the complex action look like a single overall reaction. Examples include; the bacterial glycogen debranching enzyme formed of 2 subunits each has a different enzymatic activity; bacterial acetyl-CoA carboxylase with 2 enzymatic actions but three subunits due to presence of the third biotin carrier peptide; and, the bacterial pyruvate dehydrogenase that has 5 enzyme activities in 5 subunits that are repeated to form a complex formed from 64 subunits. The functional dimeric multifunctional complex of vertebrate fatty acid synthetase is another example (Figure 3).

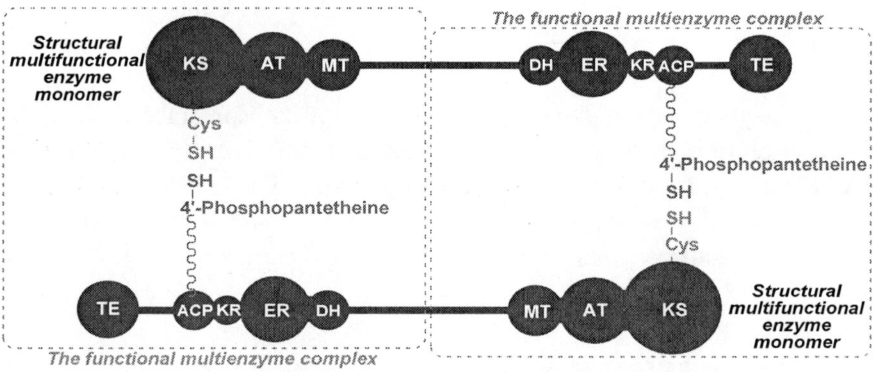

Figure 3. Domains and enzymatic activities of vertebrate fatty acid synthase multienzyme dimer. The structural subunits are identical and each monomeric multifunctional enzyme polypeptide contains all fatty acid synthase required enzymatic activities in three distinct domains. The 1st domain contains acetyl transferase (or transacylase; AT), malonyl transferase (or transacylase; MT), and ketoacyl synthase (or the condensing enzyme; KS). The 2nd domain contains the acyl carrier protein (ACP), β-ketoacyl reductase (KR), dehydratase (or hydratase, DH), and enoyl reductase (ER). The 3rd domain contains thioesterase (TE). The dimer is arranged head-to-tail so as the functional subunit will have a complete set of complementary enzymatic activities form one half of each monomer. The long and flexible 4'-

phosphopantetheinyl group bound to ACP carries the fatty acyl intermediates from one catalytic site on a functional subunit to another. The other hand of it is the -SH group of a cysteine residue of KS.

Both multienzyme complexes and multifunctional enzymes by bringing the enzymes of a pathway together in one group (such complexes are called metabolons), insure a rapid and efficient process by avoiding competing pathways for their intermediary substrate forms and by building up activating local high substrate concentrations. This is why they were adopted in biological systems. Therefore, the multienzyme complexes represent nature's adaptation to maximize the economy of the resources. Most of these complexes have a common carrier subunit that handles the intermediates and delivers them in a sequential manner to the enzymes. This kind of transfer of substrates and intermediates from one enzyme to the next is known as 'substrate channeling'. Certain enzymes of the citric acid cycle have been isolated together as supramolecular aggregates, or have been found associated with the inner mitochondrial membrane.

Other enzymes are *monofunctional single reaction enzymes* and catalyze one enzymatic reaction (with or without related reactions), e.g., digestive enzymes and carbonic anhydrase (EC4:2:1:1). They are either monomeric or multimeric in structure. Monomeric single reaction enzymes include; human succinate dehydrogenase, carbonic anhydrase II and trypsin. Multimeric single reaction enzymes include; human glucose-6-phosphate isomerase (2 subunits); glucose-6-phosphate dehydrogenase (4 subunits), and, NADH dehydrogenase (NADH-coenzyme Q reductase; 24 subunits). Like other proteins, the monomeric enzymes display primary, secondary and tertiary structures, whereas, the multimeric enzymes possess quaternary structure as well.

Substrate Specificity of Enzymes

Remember that the most striking differences between simple catalysts and enzymes are their substrate specificity and catalytic efficiency. Both are essential for the regulated metabolic activities in all biological forms of life. Substrate specificity and catalytic efficiency are dependent on the specific binding energy between chemical groups of the substrate and the chemical groups at the active sites of the enzyme. Factors that affect the catalytic ability of an enzyme may also affect its specificity.

Specificity is the ability of the enzyme to discriminate its substrate(s) from several substances in a mixture competing for its active site. An enzyme is expected - as practically proved - to possess very high maximum velocity (V_{max}) and very high affinity, i.e., low K_m for its best substrate. The number of substrate molecules handled by one active site per second is called k_{cat}. Therefore, the catalytic efficiency of an enzyme can be expressed in terms of k_{cat}/K_m ($S^{-1}M^{-1}$). This value is highest for the best substrate and is called the specificity constant and incorporates the rate constants for all steps in the reaction. Because the specificity constant reflects both affinity and catalytic ability, it is useful for comparing different enzymes against each other, or the same enzyme with different substrates.

As an applied example, the human brain hexokinase activity on different substrates will be explained. On D-glucose, it has a normalized velocity (K_{cat}) of 1.0 μmoles/second; a K_m of 1 x 10^{-4} μM/L; and, hence, its K_{cat}/K_m equals 10,000, i.e., the best specificity. On D-fructose, it has K_{cat} of 1.5 μmoles/second; K_m of 0.2 μM/L; and, hence, K_{cat}/K_m equals 7.5. On D-galactose, it has K_{cat} of 0.02 μmoles/second; K_m of 1.0 μM/L; and, hence, K_{cat}/K_m equals 0.02. On 6-deoxy-D-glucose (an expected inhibitor of the enzyme used as anticancer therapy by inhibiting the glycolytic metabolism of cancer cells), it has K_{cat} of 0.0 μmoles/second; K_m of 0.025 μM/L; and, hence, K_{cat}/K_m equals 0.0. On D-arabinose, it has K_{cat} of 0.10 μmoles/second; K_m of 25 μM/L; and, hence, K_{cat}/K_m equals 0.004. On D-xylose, it has K_{cat} of 0.0 μmoles/second; K_m of 0.167 μM/L; and, hence, K_{cat}/K_m equals 0.0.

X-ray diffraction crystallography of the enzyme-natural substrate complex revealed that the three dimensional structure of an active site is compatible with the configuration (a unique three dimensional arrangement of the atoms, bonds and bond angles of a molecule) of the ligand (substrate). For example, an enzyme that expects a chemical moiety in the substrate, e.g., an -OH with D-orientation, would simply be unable to accommodate an L-isomer with the exception of epimerases. Therefore, specificity is due to ability of an enzyme to differentiate between minor configurational differences among different substances; see also models of substrate-enzyme binding below.

One of the enzymes showing the highest specificity and accuracy are involved in the copying and expression of the genome. These enzymes have "*proof-reading*" mechanisms, i.e., an enzyme such as DNA polymerase catalyzes a reaction in a first step and then checks that the product is correct in a second step. This two-step process results in average error rates of less than $1/{\sim}10^9$ reactions in high-fidelity mammalian DNA polymerases. Similar proofreading mechanisms are also found in RNA polymerase, aminoacyl-tRNA synthetases and ribosomes.

Enzyme Substrate Specificity May be Subdivided into 5 Types:

- Stereospecificity.
- Absolute specificity.
- Dual specificity.
- Relative (broad or low) specificity.
- Structural (intermediate) specificity.

Stereospecificity:

\Enzymes are generally specific for a particular steric configuration (enantiomers, i.e., enantioselectivity) of a substrate; D- or L-isomers. Examples include; glucokinase that phosphorylates D-glucose but not other hexoses or L-glucose; fumarase interconverting fumarate (trans-) ⇔ L-malate but not maleic acid (cis-fumarate) or D-malate, and, lactate dehydrogenase that interconverts pyruvate ⇔ L-lactate but not D-lactate. Most of the body metabolic enzymes act on D-forms of sugars and L-forms of amino acids. Exception to this generalization is racemases that are enzymes reversibly interconvert the D-isomers and L-isomers of specific substrate and epimerases.

Absolute Specificity:

The enzyme acts on only one substrate, e.g., uricase enzyme acts on uric acid, arginase enzyme acts on arginine, succinate dehydrogenase interconverting succinate and fumarate, urease enzyme acts on urea, carbonic anhydrase enzyme acts on carbonic acid.

Dual Specificity:

There are 2 subtypes of dual specificity:

- *Enzyme may act on 2 different substrates to catalyze one type of reaction*, e.g., xanthine oxidase oxidizes hypoxanthine and xanthine into uric acid; Hypoxanthine ⇒ Xanthine ⇒ Uric acid.
- *Enzyme may act on one substrate to catalyze 2 different types of reactions*, e.g., isocitrate dehydrogenase that dehydrogenates and decarboxylates isocitrate into α-ketoglutarate; Isocitrate + NAD ⇒ CO_2 + NADH.H^+ + α-ketoglutarate.

Relative (broad or low) Specificity:

The enzyme acts on a group of compounds related to each other in having the same type of bond to catalyze the same type of reaction, e.g., L-amino acid oxidase acting on several L-amino acids, and, D-amino acid oxidase (EC1.4.3.3) acting several D-amino acids. Hexokinase is another example since it phosphorylates all D-hexoses. Lipase catalyzes hydrolysis of ester linkage present in triglycerides containing different types of fatty acids. Amylase catalyzes hydrolysis of glycosidic linkages present in starch, dextrin or glycogen. Proteases hydrolyze peptide bonds in different proteins. Alcohol dehydrogenase catalyzes oxidation-reduction reactions upon a number of different alcohols, ranging from methanol to butanol. However, enzymes having broad substrate specificity are generally most active against one particular substrate. Therefore, alcohol dehydrogenase has ethanol as the preferred substrate. It has been suggested that this broad substrate specificity is important for the evolution of more specific new biosynthetic pathways. It also shows that such specificity could be modulated molecularly by minor mutations in the gene of the enzyme as an approach to develop novel biocatalysts.

Structural (intermediate) Specificity:

The enzyme is specific to the bond in a group of related compounds like the relative specificity but it requires specific chemical groups or atoms around the target bond. Pepsin hydrolyzes the internal or terminal peptide linkages formed by the amino groups of phenylalanine or tyrosine. Trypsin attacks the peptide linkage containing the carbonyl group of arginine or lysine. Chymotrypsin hydrolyzes many different peptide bonds formed of a carbonyl group contributed by phenylalanine, tyrosine or tryptophan. This type of specificity is sometimes described as *group specificity* such as amino-peptidases and carboxy-peptidase, both break terminal peptide bonds of a peptide chain from its amino end or its carboxy end, respectively.

Catalytic Specificity:

Enzymes are also specific in the nature of chemical catalysis they execute. For example, glucokinase would catalyze phosphorylation but not oxidation or cleavage of glucose. However, an enzyme may have another type of what is called *side-reaction* that it also catalyzes. Example is the hepatic lipase that is mainly a lipase but it also has a phospholipase activity. Such side-reaction activity could be augmented - with and without affecting the original enzyme activity - by targeted mutagenesis and molecular evolutionary manipulations of the gene of the enzyme as a tool for design of novel biocatalysts. Both substrate and catalytic specificities reflect the specific structure-function design of the enzymes with certain binding and catalytic chemical groups at their active sites.

EXTRA ATTENTION: SHIFT OF THE ENZYME SPECIFICITY

Some enzymes change their substrate and catalytic specificity. This could be induced by several mechanisms including; i) *partial proteolytic cleavage*, e.g., NAD^+-dependent xanthine dehydrogenase (as the safe hypoxanthine/xanthine catabolizing form of xanthine oxidase) that gets proteolytically cleaved during hypoxia and cell damage into the oxidase form that uses instead O_2 as hydrogen acceptor and generates O_2 free radical (superoxide anion and H_2O_2) during reperfusion injury, ii) *binding of specific allosteric effector*, e.g., the deoxyribonucleoside diphosphate reductase that reduces pyrimidines (CDP and UDP) into dCDP and dUDP upon ATP binding, whereas, the binding of dTTP to the same enzyme induces the reduction of GDP into dGTP, that in turn induces the reduction of ADP into dADP, and, iii) *binding of a regulatory protein subunit*, e.g., lactose synthase that works as galactosyl-transferase in non-lactating mammary glands and other tissues to catalyze galactosylation reactions, and the synthesis of N-acetyl-galactosamine from UDP-galactose and N-acetyl-glucosamine. After parturition, the prolactin hormone induce synthesis of the milk α-lactalbumin. Accumulated α-lactalbumin binds the transferase as a regulatory subunit to change the enzymes specificity to be lactose synthase, iii) *Rational design and directed molecular evolutionary enzyme engineering*; See later for details. Some enzymes show tissue-specific change in their specificity though they are the same enzyme, e.g., acyl-CoA synthetase (EC 6.2.1.3) - despite the ide range of its specificity of long-chain saturated and unsaturated fatty acids - in the liver activates fatty acids with 6-20 carbons, whereas, that of the brain hold high activity towards fatty acids up to 24 carbons in chain length.

TURNOVER NUMBER (K_{CAT}) OF ENZYMES

Turnover number (K_{cat}) is the catalytic unit of the enzyme action, i.e., the maximum number of moles of substrate converted into product per mole of the enzyme catalytic site per second using a pure enzyme preparation - under substrate saturating conditions. If the molecular concentration of enzyme [E] corresponding to V_{max} is known, it can be used to calculate the value of K_{cat}, since $V_{max} = K_{cat}[E]$. Brackets [-] define concentration in moles. K_{cat} is a measure of the rate at which each enzyme molecule turns over substrate into product. Examples include; catalase (EC 1.11.1.6) with 40,000,000 molecules of H_2O_2 converted/S^{-1}, fumarase with 800 molecules of fumarate hydrated/S^{-1}, whereas, lysozyme has 0.5 hexa-N-acetylglucosamine of the bacterial wall complex carbohydrate hydrolyzed/S^{-1}. Therefore, enzymes massively differ in their catalytic efficiency, i.e., their turnover number.

Chapter IV

MECHANISM OF ENZYME CATALYSIS

BASIC COMPONENTS OF AN ENZYME CATALYZED REACTION

As explained from the general reaction $\mathbf{S} \xrightarrow{A \text{ or } I} \mathbf{P}$ are; i) S = the substrate(s); ii) P = product(s); iii) A = activator(s); and, iv) I = inhibitor(s). Substrate(s) are the reactant compound(s) of the reaction upon which the enzyme acts. Product(s) are the resultant compound(s) of the reaction. The reaction activators include the holoenzyme system (enzyme/prosthetic group, coenzyme and cofactor) and other activators including the substrate(s) themselves. Inhibitor(s) are substance(s) that reduce or block the reaction; they include the reaction product(s) themselves. *The reaction modulators* are substances that alter the rate of reaction positively (activators and substrate) or negatively (inhibitors and product). The single substrate-enzyme-catalyzed reactions have three basic steps; binding of the substrate S; conversion of the bound substrate ES into bound product EP; and, release of the product P - for enzyme recycling. Conversion of the bound substrate ES into bound product EP is the catalyzed step that may goes through several intermediary sub-steps, i.e., transition states.

Activation Energy ($\Delta G^{\ddagger}$, kJ/M):

Reactions that proceed from initial substrates to products consume energy (-ΔG) to reach their reactivated transition state. The required free energy difference between the energy levels of the substrate ground and transition states is called activation energy. Activation energy is the amount of energy required to raise all the molecules in one mole of substance(s) to the transition state and to stabilize it. At the transition state molecules are reversibly ready to split into or collide to form product(s). Activation energy is high in the non-enzyme catalyzed reaction (*uphill, over the mountain reaction*), but the presence of the substrate as enzyme-substrate complex that is composed of multiple transitional interaction highly reduces such requirement through alternative reaction route (*a tunnel through the mountain reaction*). This avoids the necessity of raising reactant temperature - as in the *in vitro* - to unphysiological limits that does not suite the fixed body temperature. However, the non-enzyme catalysis acts by reducing the required activation energy, too. Activation energy is invested in; alignment of reactive groups, formation of transient unstable charges, bond rearrangement, and other transformations.

The Free Energy and Metabolic Reactions:

The Gibbs' free energy (ΔG) of a reaction is the maximum amount of energy that can be obtained from a reaction at constant temperature and pressure. The units of free energy are kcal/mol (kJ/mol). It is not possible to measure the absolute free energy content of a non reacting substance. However, when reactant A reacts to form product B, the free energy change in this reaction ΔG, can be determined. For the reaction $A \Rightarrow B$: $\Delta G = GB - GA$; where GA and GB are the free energy of A and B, respectively. All reactions in biologic systems are considered to be reversible reactions, so that the free energy of the reverse reaction, $B \Rightarrow A$, is numerically equivalent but opposite in sign to that of the forward reaction. If the concentration of B is greater than that of A at equilibrium, it reflects the fact that the reaction $A \Rightarrow B$ is favorable to move forward from a standard state in which A and B are present at equal concentrations. Therefore such reaction is a spontaneous or exergonic reaction. The free energy of such reaction is negative and $\Delta G < 0$ because energy is liberated by the reaction. However, if the concentration of A is greater than that of B at equilibrium, it reflects the fact that it reflects the fact that the forward reaction would be unfavorable, nonspontaneous or endergonic and its free energy is positive; $\Delta G > 0$. This means that the reaction requires energy input to be pushed forward A $\Rightarrow$ B - from such equilibrium imbalance to the standard state in which A and B are present at equal concentrations. Otherwise, the backward reaction B $\Rightarrow$ A will be favored. The total free energy available from a reaction depends on both its tendency to proceed forward from the standard state (ΔG) and the amount (moles) of reactant converted to product.

Thermodynamic measurements are investigated at *standard-state conditions* where reactant and product have equimolar concentration (1 molar), the pressure of all gases is 1 atmosphere and the temperature is 25 °C (298 °K). Most commonly, the concentrations of reactants and products are then measured when the reaction reaches equilibrium. Standard free energies are represented by the symbol $\Delta G°$ and biological standard free energy change by $\Delta G°'$, with the accent symbol designating pH 7.0. The free energy available from a reaction, may be calculated from its equilibrium constant by the Gibbs equation: $\Delta G^{o'} = RT \ln K'_{eq}$; where T is absolute temperature (°Kelvin), $\ln K'_{eq}$ is the natural logarithm of the equilibrium constant for the reaction at pH 7.0, and R is the gas constant: $R = 8.3\ JK^{-1}M^{-1}$ or $-2\ cal\ K^{-1}M^{-1}$.

Example is the hydrolysis reaction of glucose-6-phosphate into glucose + phosphate (Glucose-6-P + H_2O $\Rightarrow$ glucose + phosphate). Since its free energy $\Delta G^{o'}$ is negative = -3.3 kcal/mol, it occurs spontaneously in this favored directed. However, the metabolically required reaction in the opposite direction, i.e., to synthesize glucose-6-phosphate from glucose and phosphate, would require input of energy equal $\Delta G°' = +3.3$ kcal/mol. Biologically, this reverse endergonic reaction is favored through its energetic coupling to the exergonic reaction of ATP hydrolysis with a bigger negative $\Delta G°'$ (-7.3 kcal/mol), so as the combine reaction with phosphate transfer activated by a kinase (gluco-/hexokinase) energetically favors the formation of glucose-6-phosphate with an overall exergonic negative $\Delta G°'$ (-4.0 kcal/mol). The enzyme catalysis further lowers the required energy input;

Glucose + phosphate $\Rightarrow$ glucose-6-phosphate + H_2O ($\Delta G°' = +3.3$ kcal/mol)

ATP + H_2O $\Rightarrow$ ADP + phosphate ($\Delta G°' = -7.3$ kcal/mol)

Glucose + ATP $\Rightarrow$ glucose-6-phosphate = ADP ($\Delta G°' = -4.0$ kcal/mol)

Biological systems use such coupling with ATP hydrolysis for several metabolic reactions particularly the biosynthetic ones besides the so called active processes such as transport across membrane and muscle contraction. The biochemical metabolic intermediates with a hydrolyzable bond free energy changes equal to or greater than that of ATP hydrolysis into ADP + phosphate, are called *high-energy bond containing compounds* (usually an anhydride or thioester bonds). Examples include; phosphoenol pyruvate ($\Delta G^{\circ\prime}$ = -14.8 kcal/mol), creatine phosphate ($\Delta G^{\circ\prime}$ = -12.0 kcal/mol), 1,3-diphosphoglycerate ($\Delta G^{\circ\prime}$ = -11.8 kcal/mol), pyrophosphate ($\Delta G^{\circ\prime}$ = -8.0 kcal/mol), and acetyl-CoA ($\Delta G^{\circ\prime}$ = -7.5 kcal/mol). Compounds with a hydrolyzable bond free energy changes lower than that of ATP hydrolysis into ADP + phosphate are called *lower-energy bond containing compounds* (mostly phosphate esters). Examples include; glucose-1-phosphate ($\Delta G^{\circ\prime}$ = -5.0 kcal/mol), fructose-6-phosphate ($\Delta G^{\circ\prime}$ = -3.8 kcal/mol), and glucose-6-phosphate ($\Delta G^{\circ\prime}$ = -3.3 kcal/mol). The exergonic reactions are termed *catabolism* (breakdown and/or oxidation of fuel molecules), whereas, endergonic reaction are termed *anabolism*. Catabolism and anabolism constitutes the two aspects of *metabolism*.

The reaction rate directly proportionates with % of reactant molecules already reached the transition state and inversely with the amount of required $\Delta G^{\ddagger}$. For any reaction to proceed, the energy content of the reactants must be greater than the energy content of the products. Thus, in the *exergonic reaction* with *the standard free-energy* change ($\Delta G^{o'}$, kJ/mole) is negative $-\Delta G^{o'}$, because reactant(s) liberates energy during conversion into lower energy product(s) rather than absorbing energy (as in *endergonic reaction*) and is expected to proceeds spontaneously because it is thermodynamically favorable towards product formation. The change in the free-energy (ΔG) is the driving force of any reaction, e.g., A+B $\Leftrightarrow$ C+D; the $\Delta G = \Delta G^{o'} + RT \ln [C][D]/[A][B]$. Since, $\Delta G^{o'}$ -2.303 RT $\log_{10} K'_{eq}$ is fixed for particular reaction, R is the gas constant = 1.98 x 10^{-3} (or 8.315 J/mole), T = temperature = 298 ^{o}K, and, K'_{eq} = reaction equilibrium = [C][D]/[A][B], therefore, for the K'_{eq} = 10^{-6}, the required $\Delta G^{o'}$ is 34.2 kJ/mole; for the K'_{eq} = 10^{-3}, the required $\Delta G^{o'}$ is 17.1 kJ/mole; for the K'_{eq} = 10^{-1}, the required $\Delta G^{o'}$ is 5.7 kJ/mole; for the K'_{eq} = 1, the required $\Delta G^{o'}$ is 0.0 kJ/mole; for the K'_{eq} = 10^{1}, the required $\Delta G^{o'}$ is –5.7 kJ/mole, and, for the $K_{eq'}$ = 10^{3}, the required $\Delta G^{o'}$ is –17.1 kJ/mole. The transition state theory is stating that the reaction rate constant (K) $\propto 1/\ln\Delta G^{\ddagger}$, i.e., K exponentially and inversely proportionates with $\Delta G^{\ddagger}$.

However, the negative value of $\Delta G^{o'}$ does not determine the rate of the reaction. Example, although glucose oxidation through several separate steps (glycolysis-citric acid cycle-oxidative phosphorylation) by O_2 into CO_2 and H_2O ($C_6H_{12}O_6 + 6O_2 \Rightarrow 6CO_2 + 6H_2O$) has $\Delta G^{o'} = -686$ Kcal/mole, without catalyst and by-passing the activation energy barrier it will never happen, whereas, presence of appropriate chemical or enzymatic catalysts would require seconds to accomplish it. Therefore, requirement of activation energy is essential for the stability of molecules (simple and macromolecules) in the biological system, because without it reaction would be spontaneous and freely reversible and nothing will be stable in our body. While the reaction equilibrium directly correlates $\Delta G^{o'}$, the reaction rate (*Velocity*) inversely correlates $\Delta G^{\ddagger}$ and positively proportionates with concentrations of the reactants (Figure 4).

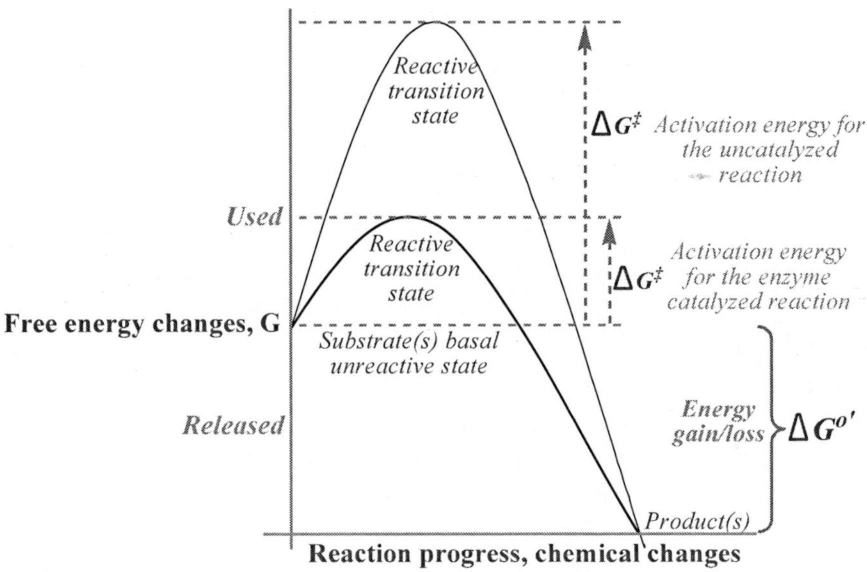

Figure 4. The role of the activation energy ($\Delta G^{\ddagger}$) and the standard free-energy ($\Delta G^{o'}$) in the enzyme catalyzed and uncatalyzed reaction.

One major factor by which the enzyme lowers $\Delta G^{\ddagger}$ is *the binding energy* (ΔG_B) released upon the formation of different bonds (covalent and non-covalent) between toms (or groups) at the active site of the enzyme and those of the substrate. The binding energy also is an essential determinant in the enzyme specificity. To accelerate a first-order reaction (see details below) by a factor of 10, the $\Delta G^{\ddagger}$ must be lowered by ~5.7 kJ/mole. A single weak bond formation between the active site and the substrate releases 4 – 30 kJ/mole. Therefore, formation of a large number of such bonding generates 60 – 100 kJ/mole that explains the speed at which the enzyme reaction is executed.

Minimal change in the target interacting groups in the substrate that may have little effect on the formation of the ES complex (as the case with alternative substrates) will not significantly affect the reaction kinetic parameters (the dissociation constant, K_d; and K_m, if $K_d = K_m$) that reflect the $E + S \Rightarrow ES$ equilibrium. However, such same change will greatly affect the overall reaction rate (k_{cat} or k_{cat}/K_m), because the bound alternative substrate lacks potential binding interactions needed to lower the activation energy. *Vice versa*, any change in the interacting groups of the active sites of the enzyme (as the case with missense mutations) will have similar consequences.

This modeling of the enzyme catalyzed reaction obeys *the laws of thermodynamics* and fits more *the induced-fitting model* of the enzyme-substrate interaction where transitional conformational changes are attained by both substrate and enzyme. In *the lock-and-key model*, although the initial binding lowers the activation energy, the rigid complex will not be able to proceed further. Like the general uses of the activation energy, the binding energy is utilized to; i) lower entropy; ii) maximize desolvation, and, iii) provide energy required for activating electron redistribution or molecular distortion.

Other way by which Enzymes lower $\Delta G^{\ddagger}$ to hasten the Reaction Rate Include;

- Lowering the activation energy by creating an environment in which the transition state is stabilized (e.g. straining the shape of a substrate into their transition state form, thereby reducing the amount of energy required to complete the transition).
- Lowering the energy of the transition state, but without distorting the substrate, by creating an environment with the opposite charge distribution to that of the transition state. Such an environment does not exist in the uncatalyzed reaction in water.
- Providing an alternative pathway by temporarily reacting with the substrate to form an intermediate ES complex that is impossible in the absence of the enzyme.

- Reducing the reaction entropy change by bringing substrates together in the correct orientation to react and destabilization of their ground state.

EXTRA ATTENTION: LAWS OF THERMODYNAMICS

The 1st law of thermodynamics or the law of conservation of energy states that the total energy of a system, including its surroundings, remains constant. This implies that a change in energy within a closed system is neither lost nor gained but it can get transferred from one part of the system to another and/or from one form to another within the system. Therefore, in a biological system chemical energy is transformed into heat, or electrical, radiant and/or mechanical energy. *The 2nd law of thermodynamics* states that the total entropy (S) of a system must increase if a process is to occur spontaneously, where; entropy is the extent of disorder or randomness of the system. Entropy is maximized in a system as the system approaches its true equilibrium. Under standard state conditions of constant temperature and pressure, the free energy change of a reacting system (ΔG) equals $\Delta H - T\Delta S$ ($\Delta G = \Delta H - T\Delta S$), where, ΔH is the change in enthalpy (heat), T is the absolute temperature, and ΔS is the change in entropy. Under the body conditions of a biochemical reaction, since ΔH approximately equals the total change of the internal energy of the reaction (ΔE), then, $\Delta G = \Delta E - T\Delta S$. Therefore, exergonic reactions will be spontaneous and will have a negative ΔG ($-\Delta G$). Moreover, a $-\Delta G$ with a great value makes the reaction essentially irreversible. Oppositely, endergonic reactions will proceed only if extra free energy could be gained and will have a positive ΔG ($+\Delta G$). A ΔG value of zero means that the system is stable at equilibrium without net change. If the standard state condition of equimolar concentrations of the reactants (1.0 mol/L each) is applied then ΔG is said to be the standard free energy change; ΔG^o. ΔG^o at the conditions of the body biochemical reactions, i.e., at pH 7.4, is denoted $\Delta G^{o'}$. Since in coupled exergonic/endergonic reactions, some of the energy liberated from one is transferred to the other and not all of the liberated energy was lost as free heat, it is not appropriate to call them exothermic and endothermic, respectively. Therefore, the terms endergonic and exergonic denotes processes of gain or loss of free energy irrespective to the form of energy involved.

ENZYME-SUBSTRATE INTERACTION

During the enzyme action, there is a temporary binding (covalent and non-covalent) between the chemical groups at the enzyme active sites and its substrate to form enzyme-substrate complex. The enzyme then strained or joined bond(s) in the substrate(s) until bond(s) ruptures to give smaller products (catabolic reaction), or collide to join small substrates together producing a lager product (anabolic reaction). The enzyme is liberated in a free-state to combine with new substrate(s) and so on. So the enzyme acts only as a catalyst for speeding the arrival at the reaction equilibrium. Nevertheless, if the equilibrium is greatly displaced in one direction, as the case of a very exergonic reaction, the reaction is effectively irreversible. Under these conditions the enzyme will, in fact, only catalyze the reaction in the thermodynamically allowed direction. Despite the reversible nature of its reaction, carbonic anhydrase working at peripheral tissues with continuous production of CO_2 favors carbonic acid formation, whereas, at the lung it favors carbonic acid breakdown into CO_2 that is continuously cleared by expiration. Therefore, availability of substrate and/or withdrawal of the products shift the reaction into essentially irreversible mode under such conditions.

$$CO_2 + H_2O \xrightarrow[\textit{At tissues}]{\textit{Carbonic anhydrase}} H_2CO_3 \xrightarrow[\textit{At lungs}]{\textit{Carbonic anhydrase}} CO_2 + H_2O$$

Over the years, two models were proposed to explain mechanism by which an enzyme discriminates its ligand. The catalytic site and/or substrate-binding site were hypothesized to be rigid or flexible. In the rigid model, the enzyme has a rigid three dimensional structure and its shape does not change upon combination with substrate. Consequently, the substrate must have a complementary rigid geometric shape and size. This model was called, *the lock and key model* proposed by Emil Fisher (1894) to describe the enzyme-substrate binding.

On the other hand, the modern practically acceptable flexible *induced-fitting model*, described by Daniel Koshland (1958) proposed induced conformational changes in the three-dimensional structure of both the enzyme and substrate (to a lesser extent) to fit one another into a complex reactive transition state(s). While the lock and key model explains enzyme specificity, it fails to explain the attainment of the transition state. Therefore, the "lock and key" model has proven inaccurate, and the induced fit model is the most currently experimentally supported and accepted enzyme-substrate-coenzyme binding manner (Figure 5).

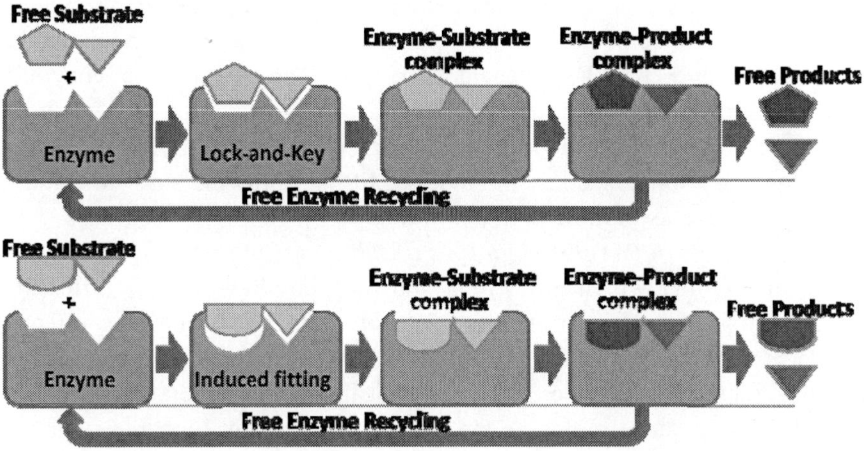

Figure 5. A simplified explanation of the enzyme-substrate binding as explained by the Lock and key (upper panel) and the Induced Fitting (lower panel) models.

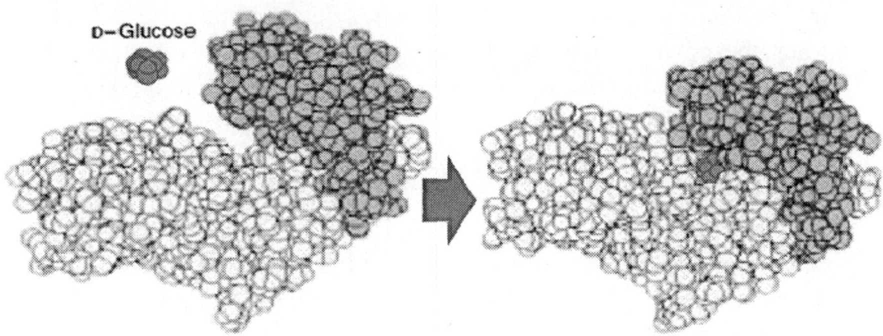

Figure 6. Conformational change in yeast hexokinase after binding to glucose as one of the largest "induced fits" known. The closed cleft forms the ATP binding site.

Understanding these models is of utmost importance in designing alternative substrates and/or inhibitors. Thus, substrate or inhibitors binding into the enzyme induce specific conformational changes in the enzyme three-dimensional conformation and volume towards the catalysis or inhibition state. The internal dynamics of enzymes is connected to their mechanism and rate of catalysis. Internal dynamics are the movement of parts of the enzyme's structure, such as individual amino acid residues, a group of amino acids, or even an entire protein domain. These movements occur at various time-scales ranging from femtoseconds to seconds. Networks of protein residues throughout an enzyme's structure can contribute to catalysis through dynamic motions. These movements are important in binding and releasing substrates and products, and may help accelerating the chemical steps in enzymatic reactions. Moreover, internal dynamics are important in understanding allosteric effects and developing new drugs. See Figure 6 for the largest known conformational changes and induced fitting following substrate binding reported for yeast hexokinase A that is similar to the human glucokinase.

There are two different mechanisms of substrate binding used by enzymes; i) uniform binding, which has strong substrate binding, and, ii) differential binding, which has strong transition state binding. The stabilizing effect of uniform binding increases both substrate and transition state binding affinity, while differential binding increases only transition state binding affinity. Both are used by enzymes to minimize the $\Delta G^{\ddagger}$ of the reaction. Enzymes which are saturated with substrate, require differential binding to reduce the $\Delta G^{\ddagger}$, whereas, the substrate-unbound enzyme may use either differential or uniform binding. However, affinity of the enzyme to the transition state through differential binding is the most important as a result from the induced fitting mechanism. The initial interaction between enzyme and substrate is relatively weak, but these weak interactions induce conformational changes in the enzyme that strengthen binding and increase the affinity to the transition state and stabilizing it. This reduces the activation energy to reach the transition state. However, the induced fitting is not a model that explains catalytic mechanisms because chemical catalysis is defined as the reduction of $\Delta G^{\ddagger}$ when the system is already in the activated ES state - relative to $\Delta G^{\ddagger}$ in the enzyme uncatalyzed reaction in water. The induced fitting only suggests that the energy barrier is lower in the closed form of the enzyme without explaining the reason for such reduction.

The formation of the enzyme-substrate complex was confirmed to exist through four approaches as follows;

- *The substrate saturation kinetics* in the presence of fixed amount of the enzyme, i.e., zero-order kinetics at V_{max} as studied by Michaelis and Menten; See below. Initially, the reaction releases a rapid burst of product nearly stoichiometric with the amount of enzyme present (in a balanced equation). The subsequent rate is slower, because enzyme turnover into its free form for new substrate acceptance is limited by the rate of the slower clearance for its transition state(s) into free product and enzyme.

- *Electron microscopy* using DNA polymerase binding on a DNA template as a model.
- *X-ray crystallography* particularly at low temperature provides excellent view of the intermediate forms of ES complex using natural substrate, substrate analogs or inhibitors.
- *Spectroscopic studies* depending on the progressive changes in absorbance and/or emission spectra upon enzyme-substrate(s) binding. Example is the fluorescence detected peaking to 500 nm for the bacterial tryptophan synthetase using L-serine and indole as substrates. The normal fluorescence of the free enzyme is intermediate between that of the form bound to L-serine (higher) and that bound to L-serine and indole (lower); see Figure 7.

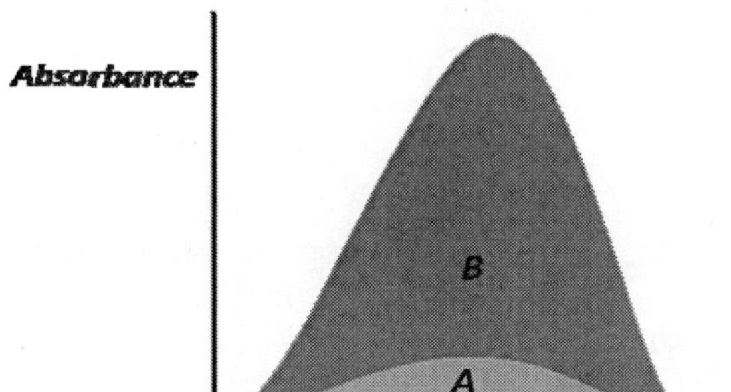

Figure 7. The absorbance spectra of tryptophan synthetase reaction in the normal enzyme free state (A) that gets higher upon L-serine binding (B) and gets lowered upon further indole into binding (C).

- *Energy diagram of the enzyme-substrate complex.* The number of steps in real enzymatic reactions results in a multi-bump energy diagram. An example is chymotrypsin reaction energy diagram initially goes down (a dip) because activation energy is provided by formation of the initial multiple weak bonds between the substrate and enzyme. As the reaction progresses, the curve raise because additional energy is required for formation of the transition state complex. This energy is provided by the subsequent steps in the reaction replacing the initial weak bonds with progressively stronger bonds. Semi-stable covalent intermediates of the reaction have lower energy levels than do the transition state complexes, and are present in the reaction energy diagram as dips in the energy curve. The final transition state complex has the highest energy level in the reaction and is therefore the most unstable state. It can collapse back to substrates or decompose to form products (Figure 8). The enzyme does not, however, change the energy levels of the substrate or product.

The transition state model was first proposed by Linus Pauling (1948) in which the enzyme is complementary in structure to the activated complexes of the reactions, i.e., to the molecular configuration that is intermediate between the reacting substances and the products of reaction. The attraction of the enzyme molecule for the activated complex would thus lead to a decrease in its energy and hence to a decrease in the energy of activation of the reaction and to an increase in the rate of reaction.

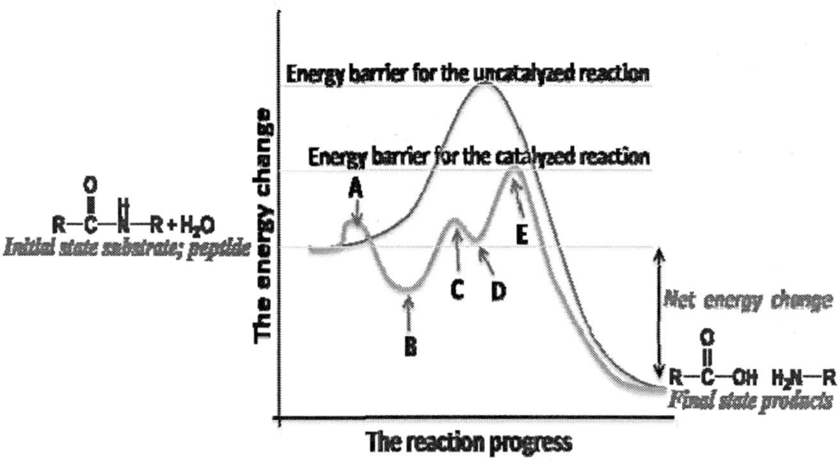

Figure 8. The postulated energy diagram for the reaction catalyzed by chymotrypsin - in the presence of enzyme (blue) and in the absence of enzyme (red). A = energy required for removing H_2O from the substrate and restricting its freedom; B = energy change after enzyme-substrate binding; C = the first enzyme-stabilized oxyanion intermediate; D = covalent-enzyme intermediate, and, E = second enzyme-stabilized transition state. The energy barrier to the transition state is lowered in the enzyme-catalyzed reaction by the formation of initial weak then final stronger bonds between the substrate and enzyme in the transition state complex.

The regulation of the rate of enzyme catalyzed reaction is achieved by controlling the quantity of the enzyme (synthesis vs. degradation), by controlling availability of substrate, activators or inhibitors, by controlling catalytic efficiency of the enzyme (e.g., by covalent modification and allosteric feedback), or optimizing the reaction temperature, pH and ionic strength (salt concentration). To study the effect of one factor of these, the other factors must be controlled constant at reaction unlimiting optimum condition for each of them.

Enzyme kinetics is the study of enzyme in action, i.e., how it binds substrate(s) and turns them into products and the mathematical studying of the rate of the enzyme catalyzed reaction and factors affecting it (positively and negatively). *Enzyme kinetics provides valuable information investable for applications that include:*

- To decipher the mechanism of the reaction.
- Determination of kinetic constants of an enzyme that reflect its mechanism, specificity and regulation.
- To obtain structural information such as: active site residues, regulatory site residues and conformational change.
- Device means to measure enzyme concentration in a biological samples for basic research and laboratory medical investigations.
- To screen for and investigate specific inhibitors that could be therapeutically invested.
- To device enzyme systems or inhibitors for useful industrial applications.

MECHANISMS CONTRIBUTING TO THE HIGH ENZYME CATALYTIC ACTIVITY

The very fast catalytic efficiency of enzyme catalyzed reaction vs. the uncatalyzed reaction (10^5–10^{17} times) under mild physiological conditions of temperature, pressure and pH in aqueous medium as compared to chemical catalysis is mainly reasoned to its ability to lower the activation energy of substrate to attain the reactive transition state and its stabilization. The initial weak conformational changes induced by substrate binding progress into stronger catalytic bindings by bringing catalytic residues in the active site close to the chemical bonds in the substrate that will be acted upon in the reaction. The high catalytic efficiency of enzymes is the corporate effect of several catalytic mechanisms working in concert within the holoenzyme system that include five possible mechanisms of "over the barrier" catalysis as well as one quantum tunneling "through the barrier" mechanism:

1. Catalysis by bond strain.
2. Proximity and orientation catalysis.
3. Acid-base catalysis.
4. Electrostatic and metal ion catalysis.
5. Covalent catalysis.
6. Quantum tunneling.

Catalysis by Bond Strain:

The affinity of the enzyme to the transition state is greater than to the substrate itself; because the induced structural rearrangements strain substrate bonds into a conformation closer to the conformation of the transition state, i.e., a ground state destabilization effect (Figure 9). Bond straining may also be induced within the enzyme itself to activate residues in the active site. This lowers the energy difference between the substrate and transition state and helps catalyze the reaction.

Substrate (Chair conformation) *Bound Substrate (sofa conformation)* *The transition state*

Figure 9. Lysozyme substrate, bound and transition state forms. On binding, the substrate conformation is distorted from the typical 'chair' hexose ring into the 'sofa' conformation, which is similar in shape to the transition state.

General acid/base Catalysis:

The functional groups of the amino acid residues (e.g., glutamate, aspartate, histidine, serine, tyrosine, cysteine, lysine and arginine) at the active site of the enzyme participate in the catalytic process as proton donors (general acids) or proton acceptors (general bases) in order to stabilize developing charges in the transition state. This typically activates nucleophile and electrophile groups, or stabilizing leaving groups. *A nucleophile* is a positive center seeking a species capable of donating electrons, e.g., O and N, whereas, *an electrophile* is a negative center seeking a species capable of accepting electrons, e.g., H. Functional groups can either be electrophilic (P in phosphate group; C in epoxides or –C=O group; and proton H^+) or nucleophilic (e.g., O in -OH, HO^- ion or $-COO^-$; N is the -N= of imidazole; S in -SH, and carboxylic groups). They constitute the two partners in the weak ionic hydrogen bond as an example of their interaction. *Acid catalysis* occurs when the partial and temporary proton transfer from an acid lowers the activation energy required to reach and stabilize the transition state and increases reaction rate. For example, the slow uncatalyzed keto-enol tautomerization due to the required high activation energy becomes faster upon proton donation to the carbonyl oxygen into the transition state that lowers the activation energy. Oppositely, *base catalysis* occurs when the partial and temporary proton abstraction by a base lowers the activation energy to reach and stabilize the transition state and increases reaction rate (Figure 10a).

Histidine is often the residue involved in these acid/base reactions, since it has a pKa close to neutral pH and can therefore both accept and donate protons. The local environment (e.g., hydrophobic environment, adjacent residues of like charge, and, salt bridge and hydrogen bond formation) of the residue substantially induces an altered pKa, to the extent that residues which are basic in solution may act as proton donors, and vice versa. Many types of biochemical reactions such as hydrolysis of a peptide bond are susceptible to general acid-base catalysis; example is the initial catalytic mechanism by serine proteases. The local environment of the bases enables histidine (an acid, pKa = 6) at the active site to accept a proton from the serine residue (a base, pKa = 14). The serine is activated into a nucleophile to attack the amide bond of the substrate (Figure 10b). The mechanism also illustrates electrostatic catalysis.

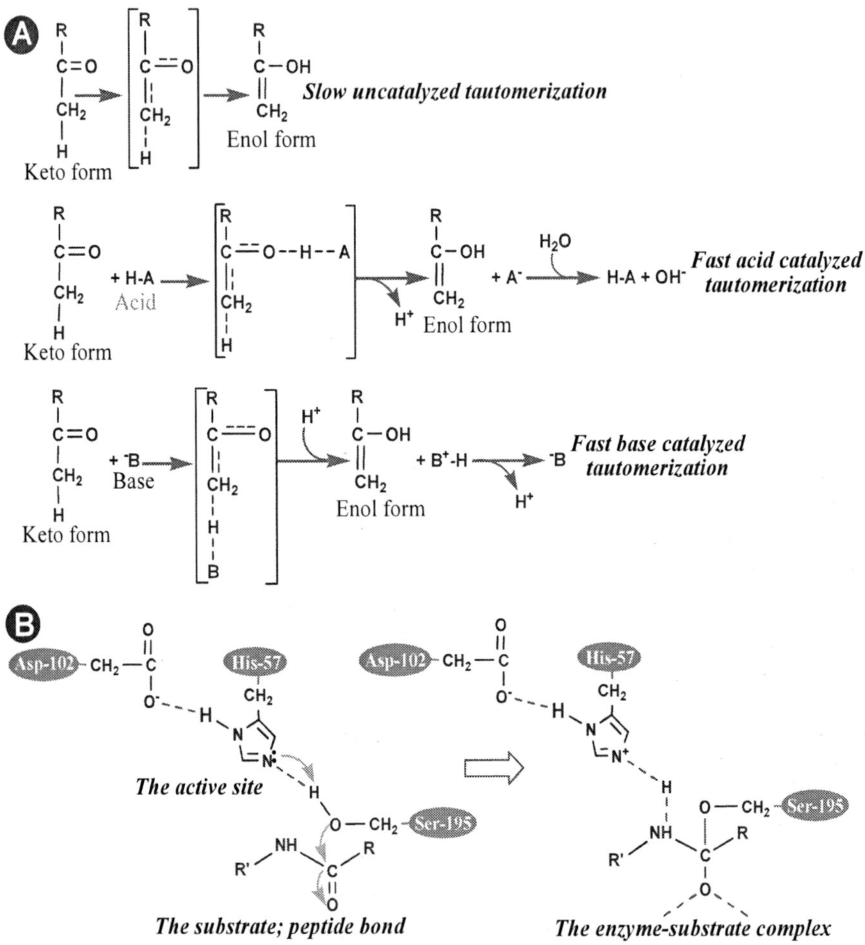

Figure 10. The general acid/base catalysis as compared to uncatalyzed tautomerization reaction (a) and optimization of the groups of the amino acid residues at the active site of serine protease (chymotrypsin) for the catalytic action (b).

Covalent Catalysis:

The transient covalent bonding of enzyme and substrate creates covalent reaction intermediate that helps reducing the energy of later transition states of the reaction and accelerate reaction rate. At a later stage in the reaction, covalent bonds are broken to regenerate the enzyme. For example, the transient covalent bonding of -OH¨ group of the activated serine residue of serine proteases and the carbonyl carbon (-C=O) of the target substrate peptide bond during the peptide bond hydrolysis by serine proteases (e.g., chymotrypsin and trypsin) into an acyl-enzyme intermediate ($-O–C-O^-$). Acidic and basic amino acid residues at active sites are readily reactive due to their charge, whereas, neutral residues require activation through interaction with neighboring residues. For instance, in serine proteases with their aspartate as the active site essential residue withdraws hydrogen from histidine, so as histidine withdraws hydrogen from serine to activate serine's -OH into alkoxide ($Ser-0^-$). Aldolase enzyme also forms Schiff base with the substrate using the free amine of its lysine residue. A second group of active site reactant (i.e., essential residue) is the functional groups of coenzymes.

Covalent catalysis occurs in two phases; i) *Phase I* cleaves the substrate to release one fragment and leaves the other fragment covalently bound to the enzyme, and, ii) *Phase II* hydrolyzes the bound fragment or transfers it into a recipient to regenerate the free enzyme. For example the cytosolic 5'-nucleotidase cleaves the nucleotide into phosphate that stays bound at its histidine essential residue and releases the nucleoside in phase I, and, then releases the phosphate in phase II. Another example is the pyruvate decarboxylase - using thiamine pyrophosphate (TPP) as the essential residue - cleaves pyruvate to release CO_2 and the remaining hydroxyethyl-TPP stays bound in phase I, and, then the hydroxyethyl-TPP is released in phase II.

Electrostatic and Metal ion Catalysis:

Electrostatic catalysis involves the acidic or basic side chains of amino acid residues (e.g., lysine, arginine, aspartic acid or glutamic acid) or metal cofactors (e.g., zinc) in the active site. They form ionic bonds or partial ionic charge interactions with the reaction intermediate. This stabilizes the charged transition states. In this concern, the metal ions are particularly effective and can reduce the pKa of water enough to make it an effective nucleophile. Electrostatic effects give the largest contribution to catalysis by enabling the enzyme to provide a microenvironment which is more polar than water.

Metal ion catalysis is utilized by about one third of all known enzymes. The required metal, e.g., Fe^{2+}, Zn^{2+}, Co^{3+} and Mn^{2+} is either a tightly bound prosthetic group as in *metalloenzymes*. Or, the element, e.g., Na^+, K^+, Mg^{2+} and Ca^{2+} may work through a loose association with the enzyme as a cofactor in *metal-activated enzymes*. These elements work either by being electron accepting Lewis acid to electrostatically stabilize or shield the negative charges, or, by being reversible electron acceptor with a change in their oxidation state to help mediating oxidation-reduction reaction.

Carbonic anhydrase as a Zn^{2+}-containing metalloenzyme catalyzes the reversible hydration of CO_2 to carbonic acid ($CO_2 + H_2O \Leftrightarrow H_2CO_3 \Leftrightarrow H^+ + HCO_3^-$). The single Zn^{2+} ion is coordinated by 3 histidine residues and a fourth hydroxyl group at the active site (histidine-$N_3 \equiv Zn^{2+}$-OH). CO_2 addition and binding at that fourth coordination position into the -OH allows its release afterwards as H_2CO_3 as the enzyme is freed to coordinate with a new -OH group.

Mg^{2+} required in the reaction of *kinases* - as metal-activated enzymes - by being electron deficient forms a complex with the phosphate groups of ATP to distorts and weakens the terminal phosphoester bond. Thus, the actual substrate for kinases is ATP-Mg^{2+} complex rather than ATP along with the phosphorylation substrate.

In the carboxypeptidase - a zinc metalloprotease - catalytic mechanism, the tetrahedral intermediate is stabilized by a partial ionic bond between the Zn^{2+} ion and the negative charge on the oxygen (Figure 11).

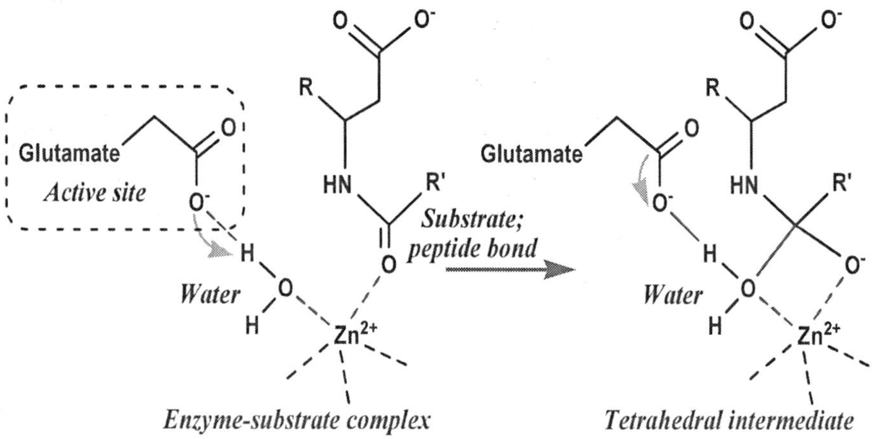

Figure 11. The electrostatic catalysis through active site glutamate and zinc ion during carboxypeptidase catalysis. The active site features allow for the activation of water that activates Zn^{2+}, and for the polarization of the peptide carbonyl group and subsequent stabilization of a tetrahedral intermediate.

Proximity and Orientation Catalysis:

The overall loss of entropy when two reactants become a single product is reduced. Therefore, optimization of the orientation of the binding groups at the active site of the enzyme to bring the susceptible bonds of the favourable substrate configuration maximizes the rate of the interaction particularly the ligations or addition reactions. Proximity is the maximization of the concentration of substrate(s) at the active site for the action of the enzyme catalytic groups. The effect is similar to an effect induced by increasing substrate concentration. Since enzymes have very high affinity (low K_m) for their substrates, they sequester substrates into their active sites (by transient covalent and noncovalent bonds). This property enables enzymes to convert intermolecular (2nd order reaction) into intramolecular (1st order) reaction. The later would be faster by hundreds of thousands of times.

Quantum tunneling:

The aforementioned traditional "over the barrier" catalytic mechanisms are different from the quantum tunneling "through the barrier" mechanisms. Some enzymes (e.g., tryptamine oxidation by aromatic amine dehydrogenase) operate with kinetics faster than predicted by the classical $\Delta G^{\ddagger}$ through tunneling protons or electrons through activation barriers. However, this mechanism functions also in uncatalyzed reactions.

SERINE PROTEASES: A MODEL MECHANISM OF CATALYTIC REACTIONS

Serine proteases (as digestive and blood clotting enzymes) hydrolyze peptide bonds and are named so for the critical catalytic serine residue at their active site. During blood clotting a peptide bond in fibrinogen is cleaved to form active fibrin by the serine protease thrombin. Thrombin has the same aspartate-histidine-serine catalytic triad found in chymotrypsin and works in essentially the same way. Thrombin is also present as an inactive zymogen precursor, prothrombin, which is itself activated through proteolytic cleavage by another blood coagulation serine protease.

The digestive serine proteases include trypsin that cleaves the peptide bond formed from carbonyl group of lysine or arginine basic amino acids, and, chymotrypsin that cleaves a peptide bond formed from carbonyl group of phenylalanine, tyrosine or tryptophan aromatic amino acids. The active sites of serine proteases have three important amino acid residues; namely, histidine, serine and aspartate. These enzymes differ in which of these residues is its binding group at the active site. This affects the conformation of the binding region in the active site for each type. In trypsin, it is a deep narrow pocket with negatively charged carboxylic group at the bottom to interact with terminal amino or imino groups of lysine or arginine. In chymotrypsin, it is a wider pocket lined with hydrophobic amino acid residues to accommodate the hydrophobic side-chains of phenylalanine, tyrosine or tryptophan. Such conformation determines the stereospecificity of each serine proteases so as D-amino acid residues will not fit into the pocket.

During chymotrypsin catalysis of the peptide bond hydrolysis, the three catalytic groups of histidine 57, aspartate 102 and serine 195 (-OH, -imidazole and HOOC-) form a catalytic triad by hydrogen bonding. In such position, serine 195 -OH proton is transferred to the histidine ring nitrogen leaving negative charge on serine oxygen to be a strong nucleophile. This transfer is facilitated and stabilized by aspartate 102 through its $-COO^-$ negative charge that stabilizes the protonation of the histidine ring. The hydrophobic aromatic ring of the residue at peptide bond to be cleaved is positioned in the hydrophobic pocket so as its carbonyl group is closer to the activated hydroxyl group of serine 195 (catalysis by proximity). The activated serine attacks the carbonyl carbon to form a tetrahedral activated transition state (see Figure 10b).

The cleavage of the peptide bond releases the C-terminal part of the polypeptide substrate by gaining the proton from histidine ring nitrogen, and, leaves the N-terminal part of the polypeptide substrate covalently bound to the enzyme - acyl-enzyme intermediate. A water molecule positioned between the acyl group and histidine 57 residue transfers one proton to the ring nitrogen of histidine 57 and its OH^- group replaces the acyl radical to form a new tetrahedral transition state. The histidine proton is transferred back to serine, and the rest of the polypeptide chain is released. The enzyme is free into its original state again; ready to catalyze a new peptide bond hydrolysis. All of these steps illustrate covalent and acid-base catalysis through proton donation and acceptance by the active groups of histidine, serine and aspartate.

Chapter V

THE FACTORS AFFECTING THE RATE OF ENZYME CATALYZED REACTION AND ENZYME KINETICS

Assuming the presence of the required coenzymes and/or cofactors, the factors affecting the maximum rate of enzyme catalyzed reaction include the following;

- Temperature.
- The pH and the ionic strength.
- Enzyme concentration.
- Substrate concentration.
- Enzyme inhibitors; including effect of radiation, light and oxidants.

To investigate the effect of each of these factors on the maximum rate of enzyme catalyzed reaction *(i.e., the velocity of the catalyzed reaction), the other factors must be controlled at their optimal limit so as the studied factor would be the only reaction rate-limiting effector.*

EFFECT OF THE TEMPERATURE ON THE RATE OF THE ENZYME CATALYZED REACTION

All enzymes work within a range of temperature specific to the organism. The velocity of chemical or enzyme reaction is almost doubled for every 10 °C increased in the reaction temperature because the increase of temperature increases proportion of substrates reaching to the readily reacting transition state. Increase of temperature increases the rate of enzyme reaction until a certain temperature at which the enzyme acts maximally beyond which rate sharply decreases because of a reduction in the enzyme activity due to its thermal denaturation and improper substrate binding and catalysis. Denaturating temperature breaks down the non-covalent interactions (hydrogen bond, ionic and hydrophobic) that stabilize the three dimensional structure of the enzyme. This derange binding and catalytic groups at the active sites of the enzyme and hence their interaction with the substrate. Oppositely, near or below the freezing temperature, the enzyme is intact but reversibly inhibited because there is no enough heat to overcome the activation energy barrier even for the catalyzed reaction.

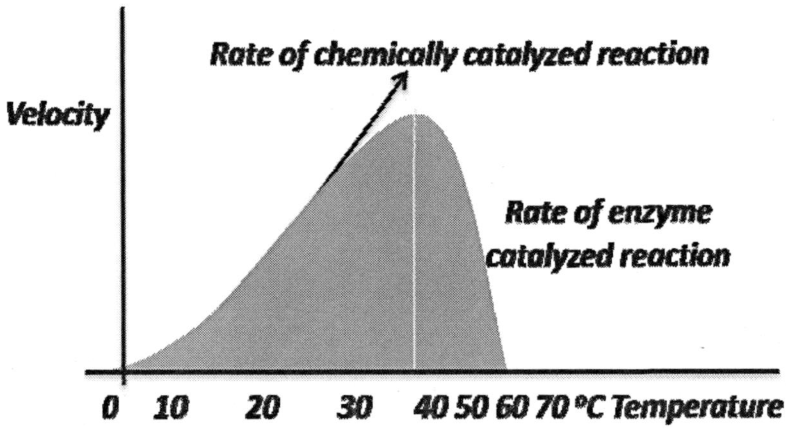

Figure 12. The effect of temperature on the rate of the enzyme catalyzed reaction as compared to the chemically catalyzed reaction.

The optimum temperature (T_{op}), ranging from 40 - 60 °C, is the one at which the rate is maximal within unlimited time because T_{op} is always a function of exposure time. T_{op} differs from one system to the other, i.e., human, cold-blooded animals, plants and thermophilic organisms. For most human enzymes T_{op} is 35 - 40 °C (average 37 °C) and for most plant enzymes T_{op} is 65 - 70 °C, whereas, enzymes in the thermophilic organisms may have T_{op} as high as 72 °C and are stable up to 100 °C (e.g., Taq DNA polymerase from Thermophilus aquaticus). Human enzymes start to denature quickly at temperatures above 40 °C. Human febrile conditions above 40 °C or freezing are fetal mainly because of the temperature effect on enzymes and other temperature-sensitive proteins. Therefore, T_{op} of an enzyme is natural temperature of the owner's organism. Oppositely, chemically catalyzed reaction would accelerate with increases in temperature in a linear fashion (Figure 12).

EFFECT OF THE PH ON THE RATE OF THE ENZYME CATALYZED REACTION

Most enzymes are sensitive to changes in the pH. Starting from a lower pH, increasing the pH leads to an increase in rate of enzyme catalyzed reaction till reaching a specific pH that is called optimum pH (pH_{op}) at which the enzyme acts maximally. Progressive increases beyond pH_{op} decrease the activity making the pH-reaction rate relationship bell-shaped. Such effect is due to the pH effect on the ionization state of amino acid residues particularly at the active sites of the enzyme and also the ionization state of the substrate. Both effects interfere with substrate binding and catalytic activity of the enzyme.

Extreme pH provided by strong acids or strong alkalis disrupts the binding forces of the enzyme three-dimensional structure (tertiary and quaternary) and may irreversibly denature it. Most of the body enzymes have pH_{op} within the physiological pH 6 – 8. However, some digestive enzymes, e.g., pepsin work at the extreme of gastric pH 1.5 – 2, and, the alkaline phosphate with pH_{op} of 9.8. Other examples include; salivary amylase that has pH_{op} of 6.8-7.0, pancreatic amylase with pH_{op} of 7.2, glucose-6-phosphatase (EC 3.1.3.9) in the cytoplasm has pH_{op} of 8.0 (Figure 13). Therapeutically administered pancreatic digestive enzymes, e.g., for patients with cystic fibrosis and several other drugs require special pharmaceutical formulation to protect them from inactivation during passage through the low pH of the stomach.

The pH_{op} in any biological system is provided by its natural buffer systems. In the *in vitro* enzymatic investigations such pH_{op} is provided by a selected buffer system. However, not only the pH_{op} is required but also *the ionic concentration of such buffer system* is critical. For instance the activity of an enzyme might be higher in 50 mM citrate buffer than 50 mM acetate buffer of the same pH. This rate difference may be attributed to; i) specific catalytic role of citrate, ii) chelation of an inhibitor by citrate, or, iii) ionic strength *per se* – since ordinary chemical reactions and likewise the rate of enzymatic reactions depend on ionic strength. Ionic strength = $0.5 \Sigma M_iZ_i^2$, where, M is the molality and Z is the charge of an ion ($_i$). Most enzymes cannot tolerate extremely high salt concentrations. The ions interfere with the weak ionic bonds of proteins. Typical enzymes are active in salt concentrations of 1-500 mM with exceptions such as the halophilic (salt loving) algae and bacteria.

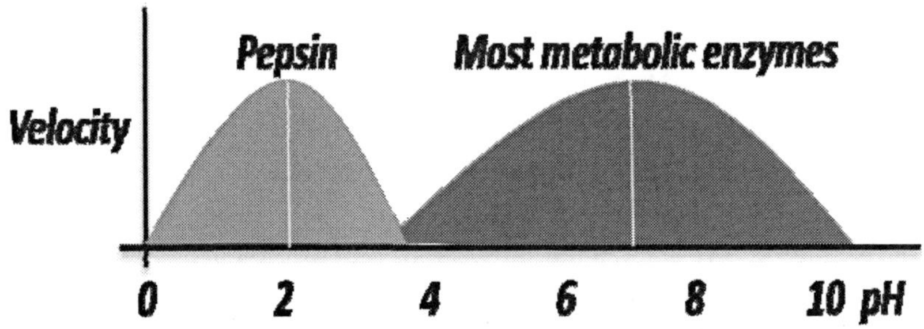

Figure 13. The effect of the change in the pH on the rate of the enzyme catalyzed reaction by pepsin as compared to the most of the intermediary metabolic enzymes.

Effect of the Enzyme Concentration on the Rate of the Enzyme Catalyzed Reaction

Alterations in the rate of a reaction are directly dependent upon the concentration of functional enzyme molecules only when the enzyme is the limiting factor in the reaction. With unlimited amount of substrate and fixed enzyme concentration, the enzyme will be saturated at V_{max} and is the rate-limiting factor.

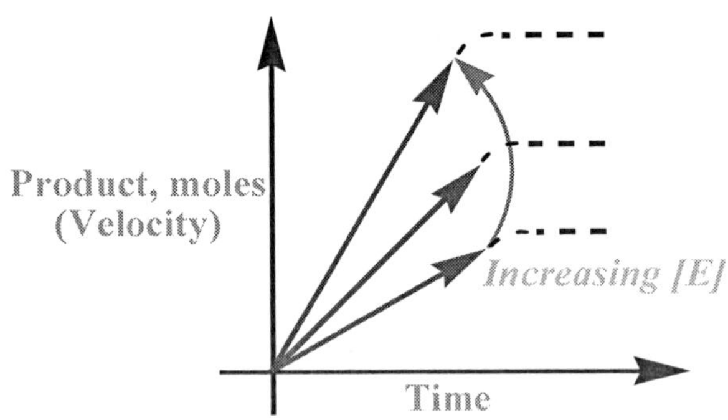

Figure 14. The effect of the change in the enzyme concentration on the rate of the enzyme catalyzed reaction.

Therefore, with increasing amount of the enzyme, V_{max} would be a function of the amount of enzyme available. With excess constant amount of the substrate and provided that the product is withdrawn, the velocities of the reaction will increase linearly with increasing concentration of the enzyme (Figure 14).

Effect of the Substrate Concentration on the Rate of the Enzyme Catalyzed Reaction

At optimal conditions, no inhibitors and a constant enzyme concentration, as the substrate (one substrate if there are several) concentration increases, the initial reaction velocity (V_o) increases gradually towards V_{max}. Plotting the changes in V_o vs. progressive increases in the substrate concentration [S] gives *the substrate saturation curve* for any enzyme catalyzing with a single substrate (Figure 15). The speed V defines the number of reactions per second that are catalyzed by an enzyme. [S] is difficult to be measured at V_{max} but rather it is measurable at 1/2 V_{max}.

$$E + S \underset{K_{-1}}{\overset{K_1}{\rightleftharpoons}} ES \underset{K_{-2}}{\overset{K_2}{\rightleftharpoons}} E + P$$

Figure 15. The effect of the change in the substrate concentration on the rate of the enzyme catalyzed reaction.

With fixed amount of the enzyme, the available substrate binding/active sites are limited. Initially these sites are all free and available for the catalytic process and the initial velocity increases in nearly linear manner with increases in substrate concentration. With further increases in substrate concentration, the rate progresses asymptotically (i.e., decelerates) till reaching the limiting velocity, i.e., maximum velocity. This is due to substrate occupancy of all the available active sites of the enzyme. Thus, there is no further increase in the reaction rate but rather a fixed amount of products is replaced by new substrate molecules at the maximum velocity. The plot of activity (rate of enzymatic reaction) vs. substrate concentration is a hyperbolic curve.

Assuming a fast reversible binding of the enzyme (E) to the substrate (S); (E + S $\Leftrightarrow$ ES) that slowly and reversibly gives rise to a free enzyme and a product (P), thus, the second reaction (ES $\Rightarrow$ P) is the rate-limiting step. The overall rate of enzyme-catalyzed reaction is proportional to the concentration of ES complex and its rate will be maximal when all the enzyme molecules are present as ES complex. The steady state maximum velocity is kept constant by replacing released products with new substrate molecules.

The kinetic mathematical studying of the substrate-reaction velocity relationship was treated through several approaches including the rapid equilibrium approach proposed by *Leonor Michaelis* and *Maud Menten* (1913). In 1902 Victor Henri proposed a quantitative theory of enzyme kinetics that did not appreciate the role of the hydrogen ion concentration. After defining the logarithmic pH-scale and the concept of buffering by P. L. Sorensen in 1909, Leonor Michaelis and his post-doctor Maud Menten confirmed Henri's equation named as Henri-Michaelis-Menten kinetics (or more commonly Michaelis-Menten kinetics); still widely used today.

Form the hypothetical reaction equation;

$$E + S \underset{K_{-1}}{\overset{K_1}{\rightleftharpoons}} ES \underset{K_{-2}}{\overset{K_2}{\rightleftharpoons}} E + P$$

and at initial reaction time, P concentration is negligible and hence K_2. Therefore, the reaction equation at that state becomes

$$E + S \underset{K_{-1}}{\overset{K_1}{\rightleftharpoons}} ES$$

The mathematical equation explaining the qualitative relationship between the substrate concentration and the rate of enzyme-catalyzed reactions - assuming that; 1) enzyme-substrate complex formation is the necessary step, 2) the rate of enzyme catalyzed reactions is determined by the rate of conversion of enzyme-substrate complex to the product and free enzyme, 3) a single substrate yields a single product, and, 4) a limited enzyme but an excess substrate - is called *Michaelis-Menten equation* that is;

$$V_o = \frac{V_{max}[S]}{K_m + [S]}$$

where V_o is the initial velocity; V_{max} is the maximum velocity (rate) that a given amount of an enzyme can attain, [S] is substrate concentration and K_m is *Michaelis-Menten constant.* Confirmation of the dependence of the initial velocity (V_o) on substrate concentration and the K_m value (substrate concentration required to produce half-maximum velocity) is provided by studying shifts in Michaelis-Menten equation in three different conditions:

- At a very low substrate concentration (much less than the K_m value).
- At high substrate concentration (much greater than K_m value).
- At substrate concentration [S] equal to K_m value.

At low substrate concentration much less than the K_m value, [S] is negligible in the equation denominator and both V_{max} and K_m as constants are replace by one constant, K. Thus;

$$V_o = \frac{V_{max}[S]}{K_m + \cancel{[S]}} = \frac{V_{max}[S]}{K_m} = \frac{\overset{K}{\cancel{V_{max}}}}{\cancel{K_m}}[S] = K[S]$$

and, thus, the initial velocity [V_o] is proportional with substrate concentration [S].

At high substrate concentration much greater than K_m value, K_m value is negligible in the denominator. Thus;

$$V_o = \frac{V_{max}[S]}{\cancel{K_m} + [S]} = \frac{V_{max}\cancel{[S]}}{\cancel{[S]}} = V_{max}$$

At substrate concentration equal to the K_m value, replacing [S] for K_m make the equation,

$$V_o = \frac{V_{max}[S]}{\underset{[S]}{\cancel{K_m}} + [S]} = \frac{V_{max}[S]}{[S] + [S]} = \frac{V_{max}\cancel{[S]}}{2\cancel{[S]}} = \frac{V_{max}}{2} = 1/2\ V_{max}$$

and when the substrate concentration is equal to the K_m value, the initial velocity is equal to half the maximum velocity. V_{max} and K_m are unique for each enzyme under its optimum pH and temperature. V_{max} reflects the catalytic efficiency of the enzyme (i.e., it proportionates with the enzyme concentration but it is not a characteristic of the enzyme), whereas, K_m reflects enzyme-substrate binding affinity and is a characteristic of the enzyme.

K_m, Michaelis-Menten constant, is the concentration of a substrate at which the catalytic activity of the enzyme reaches one-half its maximum velocity. Low K_m value for a given substrate indicates high enzyme substrate binding affinity and vice versa. Therefore, low affinity indicates lowered stability of the initial ES complex and requires higher substrate amount, whereas, high affinity requires lower amount of substrate concentration to attain half the maximum velocity. K_m is independent of enzyme concentration. V_o and [S] are measurable quantities and from the curve V_{max} and K_m for a specific enzyme can be determined. For example, hexokinase has a high glucose affinity (low K_m), whereas, glucokinase has a low affinity for glucose (high K_m). This is of utmost physiological differential regulatory importance during glucose metabolism (Figure 16). K_m is expressed in the same units as [S], i.e., mole/L. Substrates are usually present in physiological fluids at a concentration ranging around the K_m values. K_m also reflects the presence or absence of an enzyme inhibitor and its nature.

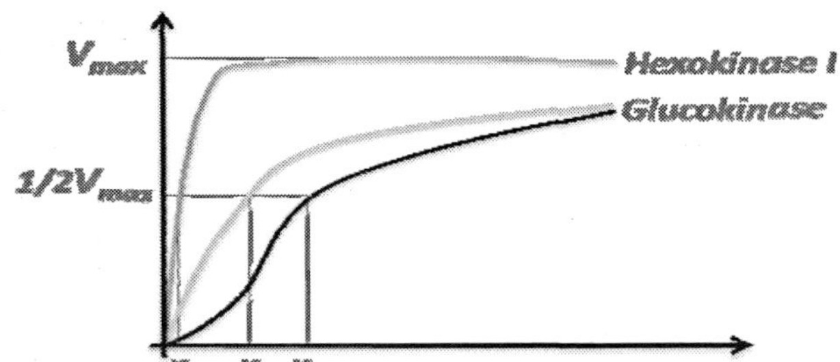

Figure 16. The catalytic difference between the RBCs hexokinase I and the liver glucokinase isoenzymes. The high affinity of hexokinase is reflected on the hyperbolic curve with very low K_m (0.05 mM glucose; blue curve). The sigmoid nature of the glucokinase curve does not obey Michaelis-Menten kinetics (possibly because the rate of an intermediate step is so slow) and reflects low affinity to glucose particularly below 5 mM (black curve; $K_{0.5}$ is 6.7 mM). Once above 5 mM glucose, the curve tends to attain nearly Michaelis-Menten kinetics (green curve).

The Linearization of Michaelis-Menten Equation:

Because of its nature as a curve that requires large number of points to construct, it is difficult to determine V_{max} and K_m from the V_o/[S] curve. It would be simpler to convert the relationship into a linear one by plotting the reciprocal values of both V_{max} and K_m ($1/V_{max}$ and $1/K_m$) and the plot obtained is called double-reciprocal plot transformation of Michaelis-Menten equation. One of these transformations is called Lineweaver-Burk plot. V_o and [S] are measurable quantities and from the reciprocal curve V_{max} and K_m for a specific enzyme can be determined (Figure 1).

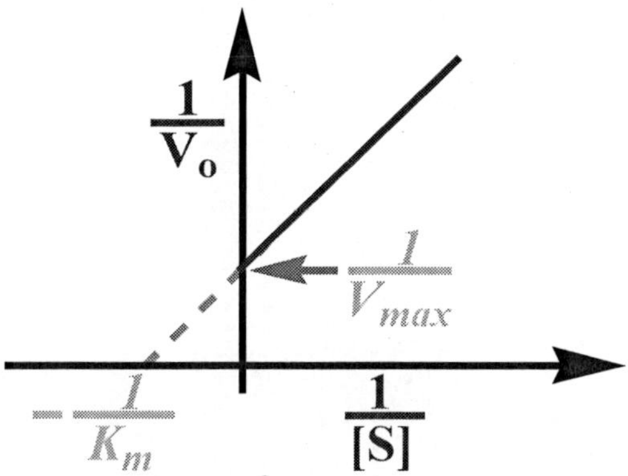

Figure 17. The Lineweaver-Burk linearization of the Michaelis-Menten substrate concentration-reaction rate relationship.

The Michaelis-Menten equation is written in reciprocal form as;

$$\frac{1}{V_o}=\frac{K_m+[S]}{V_{max}[S]}=\frac{[\not{S}]}{V_{max}[\not{S}]}+\frac{K_m}{V_{max}[S]}=\frac{1}{V_{max}}+\frac{K_m}{V_{max}}X\frac{1}{[S]}$$

In the straight line equation ($y = ax + b$), $y = 1/V_o$ and $x = 1/[S]$; plotting $1/V_o$ vs. $1/[S]$ give a straight line with the y intercept at $1/V_{max}$, x intercept at -$1/K_m$ and the line slope is K_m/V_{max}.

If $1/V_o$ in the above double-reciprocal equation is 0,

$$0=\frac{1}{V_{max}}+\frac{K_m}{V_{max}}X\frac{1}{[S]}$$

then;

$$-\frac{K_m}{V_{max}}X\frac{1}{[S]}=\frac{1}{V_{max}}$$

and multiplying both sides by

$$\frac{V_{max}}{K_m}$$

then;

$$\frac{V_{max}}{K_m} X - \frac{K_m}{V_{max}} X \frac{1}{[S]} = \frac{1}{V_{max}} X \frac{V_{max}}{K_m}$$

and therefore;

$$-\frac{1}{[S]} = \frac{1}{K_m}$$

Or,

$$\frac{1}{[S]} = -\frac{1}{K_m}$$

Thus, K_m is the negative reciprocal of 1/[S] at $1/V_o$ of zero and $1/V_{max}$ is the intercept from 1/V versus 1/S plot at 1/[S] of zero.

The theoretical maximum for the specificity constant is called the diffusion limit and is about 10^8 - 10^9 ($S^{-1}M^{-1}$). At this point every collision of the enzyme with its substrate will result in catalysis, and the rate of product formation is not limited by the reaction rate but by the diffusion rate. Enzymes with this property are called catalytically perfect or kinetically perfect. Examples of such enzymes are triose-phosphate isomerase, carbonic anhydrase, acetylcholine esterase, catalase, fumarase, β-lactamase, and superoxide dismutase.

Michaelis-Menten kinetics relies on the law of mass action, which is derived from the assumptions of free diffusion and thermodynamically-driven random collision. However, many biochemical or cellular processes deviate significantly from these conditions, because of macromolecular crowding, phase-separation of the enzyme/substrate/product, or one or two-dimensional molecular movement. In these situations, fractal (i.e., imperfect) Michaelis-Menten kinetics may be applied.

Some enzymes operate with kinetics faster than diffusion rates. Because this seems impossible, several mechanisms have been raised to explain this phenomenon. Some proteins are believed to accelerate catalysis by drawing their substrate in and pre-orienting them by using dipolar electric fields. Other models invoke a quantum-mechanical tunneling explanation, whereby a proton or an electron can tunnel through activation barriers. This suggests that enzyme catalysis may be more accurately characterized as "*through the barrier*" rather than the traditional model, which requires substrates to go "*over*" a lowered energy barrier.

Kinetics of the two-substrate Enzyme Catalyzed Reactions:

When one enzyme binds several substrates, the K_m value for each substrate reflects difference in enzyme affinity for different substrates. In reality, most enzymes work on more than one substrate, e.g., $A + B \Leftrightarrow C + D$. The enzyme-substrate complex is a ternary complex ($E + A + B \Leftrightarrow EAB \Rightarrow E + P1 + P2$) or sequential ($E + A \Leftrightarrow E' + C$, $E' + B \Rightarrow E + D$). Although the reaction equations are too complex, the K_m for each substrate and V_{max} for the reaction can be calculated from the plot of reaction kinetics. This is done by fixing one substrate at its saturating concentration and using varying concentration of the other. Isomerases are the only true one-substrate enzymes that may also include hydrolases since one substrate, i.e., H_2O is present at a constant concentration. Multi-substrate enzymes are so many an include oxidoreductases, transferases and ligases.

THE RATE AND KINETIC ORDER OF THE CHEMICAL REACTIONS

The reaction rate, i.e., the number of molecules of reactant(s) that are converted into product(s) in a specified time period depends on the reactant(s)/product(s) concentration and on the rate constants of their consumption/formation. Therefore, in $A \Rightarrow B$ reaction, the rate of the reaction equals the rate of disappearance of A, i.e., -[A], or, equals the rate of formation of B, i.e., [B]. Since concentrations of each A and B proportionate interdependently, thus, $-[A] = k[B]$ and $[B] = k[A]$, where, the k is the proportionality or rate constant. This constant is a characteristic value for every chemical reaction and is directly related to the equilibrium constant for that reaction. The rate constant for the forward reaction is defined as k_{+1} and for the reverse reaction as k_{-1}. At equilibrium, the rate (V) of the forward

A⇒B reaction equals the rate (V) of the reverse or backward B⇒A reaction, i.e., $V_{forward} = V_{reverse.}$ Since, $V_{forward} = k_{+1}[A]$, and, $V_{reverse} = k_{-1}[B]$, thus, $k_{+1}[A] = k_{-1}[B]$, i.e., $[B]/[A] = k_{+1}/k_{-1} = K_{eq}$, where K_{eq} is the equilibrium constant of the whole reaction. Thus, the equilibrium constant for a chemical reaction equals the equilibrium ratio of product and reactant concentrations, and also equals the ratio of the characteristic rate constants of the reaction.

Chemical Reactions are Classified into first Order, second Order and Zero-order Reaction Kinetics

The reaction order depends on the number of molecules involved in forming the product(s)-forming reaction complex (ES, ES_1S_2, etc). It equals the summation value of the exponents of each concentration term of reactants in the reaction rate equation. Therefore, a reaction is first order when converts the substrate A into the product B through ES without any influences from other reactants and/or solvent; since the exponent on the substrate A concentration is 1. In this case, the reaction rate proportionates with substrate A concentration - typically at the initial nearly linear portion of the reaction curve. Whereas, a reaction with two substrates (or two molecules of the same substrate) being converted into products through ES_1S_2 is a second order reaction, e.g., formation of ATP and water from ADP and phosphate. The summation of the exponents on ADP concentration (is 1) plus that of phosphate (is 1) is 2. The reaction rate proportionates with the square of the concentration, i.e., $[S] \times [S] = [S]^2$. If the reaction rate does not increase with the increase in the concentration of the substrate, e.g., when the reaction reaches the V_{max}; the reaction is said to have zero-order kinetics.

Chapter VI

ENZYME INHIBITION

INTRODUCTION

The enzyme inhibitors are low molecular weight chemical compounds that are able to reduce or completely inhibit the enzyme activity reversibly or permanently (irreversibly). Many drugs are chosen and/or designed to inhibit specific enzymes and the toxic effect of many toxins is mainly due to their enzyme inhibitory action. Therefore, studying the aforementioned enzyme kinetics and structure-function relationship is vital to understanding kinetics of enzyme inhibition that in turn is fundamental to the modern design of pharmaceuticals and industrials. Studying the enzyme inhibition kinetics and inhibitor structure-function relationship would clarify mechanisms of action and physiological regulation of metabolic enzymes; drug and toxin action and/or drug design for therapeutic uses, e.g., methotrexate in cancer chemotherapy through semi-selectively inhibiting DNA synthesis of malignant cells; the use of aspirin to inhibit the synthesis of the proinflammatory prostaglandins, and, the use of sulfa drugs to inhibit the folic acid synthesis essential for growth of pathogenic bacteria. Many life-threatening poisons, e.g., cyanide, carbon monoxide and polychlorinated biphenols are enzyme inhibitors.

Enzyme inhibitors could be classified into non-specific inhibitors and specific inhibitors. Non-specific irreversible non-competitive inhibitors include all protein denaturating factors (physical and chemical denaturation factors). The specific inhibitors attack a specific component of the holoenzyme system and maybe reversible (*by other means than increasing the substrate concentration*) or irreversible in their action. Specific inhibitors include; *1) coenzyme inhibitors*: e.g., cyanide, hydrazine and hydroxylamine that inhibit pyridoxal phosphate, and, dicumarol that is a competitive antagonist for vitamin K; *2) inhibitors of specific ion cofactor*: e.g., fluoride that chelates Mg^{2+} of enolase enzyme; *3) prosthetic group inhibitors*: e.g., cyanide that inhibits the heme prosthetic group of cytochrome oxidase; and, *4) apoenzyme inhibitors* that attack the apoenzyme component of the holoenzyme.

The apoenzyme inhibitors are of two types; i) *Reversible inhibitors;* their inhibitory action is reversible because they make reversible association with the enzyme, and, ii) *Irreversible inhibitors;* because they make inactivating irreversible covalent modification of an essential residue of the enzyme. Depending on their effect on K_m and V_{max}, the reversible apoenzyme inhibitors - also called metabolic antagonists - are of three subtypes; *a) competitive*, *b) uncompetitive* and *c) non-competitive* (or mixed type).

Extra Attention: Drug Design of Enzyme Inhibitors

Discover of useful new enzyme inhibitors used to be done by trial and error through screening huge libraries of compounds against a target enzyme. This is still successfully in use particularly compound with combinatorial chemistry approaches and high-throughput screening technology. However, rational drug design as an alternative approach uses the three-dimensional structure of an enzyme's active site or transition-state conformation to predict which molecules might be inhibitors. This shortens the screening list towards a novel inhibitor which is subsequently kinetically characterized allowing structural changes to be made to the inhibitor to optimize its binding. Alternatively, molecular docking and molecular mechanics are computer-based methods that predict the affinity of an inhibitor for an enzyme.

Irreversible Inhibition

The irreversible apoenzyme inhibitors have no structural relationship to the substrate and bind mainly covalently but also stable non-covalently with or destroy an essential functional group at the active site of the enzyme. Therefore, irreversible inhibitors may be used to identify functional groups of the enzyme active sites at which they bind. Although they have limited therapeutic application because they are usually considered to be poisons, a subset of irreversible inhibitors called *suicide irreversible inhibitors* are relatively unreactive compounds and get activated upon binding to the active site of a specific enzyme. After such binding, the suicide irreversible inhibitor is activated by the first few intermediary chemical steps of the reaction - like the normal substrate. However, it does not release as product because of its irreversible binding at the enzyme active site. Since they make use of the normal enzyme reaction mechanism to get activated and subsequently inactivate the enzyme, suicide irreversible inhibitors are also called *mechanism-based inactivators* or *transition state analog inhibitors*. This ensures that the inhibitor exploits the transition state stabilizing effect of the enzyme, resulting in a better binding affinity (lower K_i) than substrate-based designs. An example of such a transition state inhibitor is active form of the antiviral drug oseltamivir (Tamiflu; see Figure 18); this drug mimics the planar nature of the ring oxonium ion in the reaction of the viral enzyme neuraminidase. After activation in the liver, the drug replaces sialic acid as the normal substrate found on the surface proteins of normal host cells. This prevents the release of new viral particles from infected cells. It has been used to treat and prevent Influenza virus A and Influenza virus B infections. Most of these inhibitors are classified as tight-binding competitive inhibitors in other references of enzymes. However, their reaction kinetics is essentially irreversible.

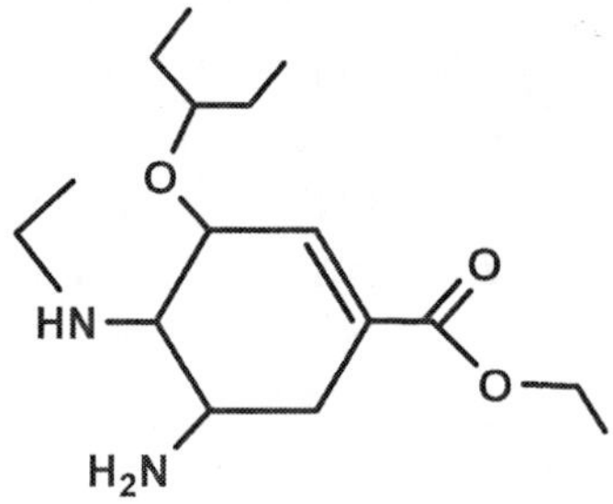

Figure 18. The transition state analog oseltamivir - the viral neuraminidase inhibitor.

The current rational drug design of new drugs is based in part on suicidal irreversible inhibitors. Chemicals are synthesized based on knowledge of substrate binding and reaction mechanisms to inhibit a specific enzyme with minimal side-effects due to non-specific binding. Transition state analogs are extremely potent and specific inhibitors of enzymes because they have higher affinity and stronger binding to the active site of the target enzyme than the natural substrates or products. However, it is difficult to exactly design drugs that precisely mimic the transition state because of its highly unstable structure in the free-state. However, it is possible to design drugs (prodrugs) that undergo initial reaction(s) to attain an overall electrostatic and three-dimensional intermediate transition state complex form with close similarity to that of the substrate. Also, the drugs could be designed to be almost like the transition state but have a stable modification; or, using the transition state analog to design a complementary catalytic antibody; called *Abzyme*.

Extra Attention: Abzymes (Catalytic Antibodies)

They are antibodies generated against analogs of the transition state complex of a specific chemical. The arrangement of amino acid side chains at the abzyme variable regions is similar to the active site of the enzyme in the transition state and work as artificial enzymes. For example, an abzyme was developed against analogs of the transition state complex of cocaine esterase, the enzyme that degrades cocaine in the body. Thus, this abzyme has similar esterase activity that is used as injection drug to rapidly destroy cocaine in the blood of addicted individuals to decreasing their dependence on it.

The saliva of leeches and other blood-sucking organisms contain the anticoagulant hirudin that irreversibly inhibits thrombin, and, to regain thrombin action synthesis of new thrombin molecules is required. This made it unsafe as an anticoagulation drug. But based on hirudin structure, rational drug design synthesized 20-amino acids peptide known as bivalirudin that is safe for long-term use because of its reversible effects on thrombin; despite its high binding affinity and specificity for thrombin.

Examples include the inhibition of ornithine decarboxylase by difluoromethylornithine that is used to treat African trypanosomiasis (sleeping sickness). The enzyme initially decarboxylates difluoromethylornithine instead of ornithine and releases a fluorine atom, leaving the rest of the molecule as a highly electrophilic conjugated imine. The later reacts with either a cysteine or lysine residue in the active site to irreversibly inactivate the enzyme. Another example is the inhibition of thymidylate synthase by fluoro-dUMP. Imidazole antimycotic drugs are examples of such group that inhibit several subtypes of cytochrome P450. The mechanisms of toxicities and antidotes of irreversible inhibitors are of medical pathological importance. Because of the irreversible inactivation of the enzyme, irreversible inhibition is of long duration in the biological system because reversal of their action requires synthesis of new enzyme molecules at the enzyme gene-transcription-translation level.

Other examples include the inhibition of acetylcholine esterase (ACE) by diisopropylfluorophosphate (DPFP), the ancestor of current organophosphorus nerve gases (e.g., Sarin and Tabun) and other organophosphorus toxins (e.g., the insecticides Malathion and Parathion and chlorpyrifos). ACE hydrolyzes the acetylcholine into acetate and choline to terminate the transmission of the neural signal form the neuromuscular excitatory acetylcholine presynaptic cell to somatic neuromuscular junction (Figure 19). DPFP as a potent neurotoxin inhibits ACE and acetylcholine hydrolysis. Failure of hydrolysis leads to persistent acetylcholine excitatory state and improper vital function particularly respiratory muscles that may lead to suffocation; with a lethal dose of less than 100 mg. DPFP inhibits other enzymes with the reactive serine residue at the active site, e.g., serine proteases such as trypsin and chymotrypsin, but the inhibition is not as lethal as that of acetylcholine esterase. Similar to DPFP, malaoxon the toxic reactive derivative from Malathion (after its metabolism by the liver) binds initially reversibly and then irreversibly (after dealkylation of the inhibitor) to the active site serine and inactivates ACE and other enzymes. Lethal doses of oral Malathion are estimated at 1 g/kg of body weight for humans.

Inhibition of ACE by these poisons leads to accumulation of acetylcholine that over-stimulates the autonomic nervous system (including heart, blood vessels, and glands), thereby accounting for the poisoning symptoms of vomiting, abdominal cramps, nausea, salivation, and sweating. Acetylcholine is also a neurotransmitter for the somatic motor nervous system, where its accumulation resulted in poisoning symptom of involuntary muscle twitching (muscle fasciculation), convulsions, respiratory failure and coma. Intoxication of Malathion is treated by the antidote drug Oxime that reactivates the acetylcholine esterase and by intravenous injection of the anticholinergic (antimuscarinic) drug atropine to antagonize the action of the excessive amounts of acetylcholine.

Another example of irreversible inhibition is iodoacetate inhibition of the glycolytic glyceraldehyde-3-phosphate dehydrogenase (GPD). Iodoacetate is a sulfhydryl compound that covalently alkylates and blocks the sulfhydryl group at the active site of the enzyme. Iodoacetate also inhibits other enzymes with -SH at the active site (Figure 20).

Allopurinol - the anti-gout drug - is a suicidal irreversible mechanism-based inhibitor of the enzyme xanthine oxidase that works as oxidase or dehydrogenase. The enzyme commits suicide by initial activating allopurinol into a transition state analog - oxypurinol - that bind very tightly to molybdenum-sulfide (Mo-S) complex at the active site (Figure 21). This enzyme accounts for the human dietary requirement for the trace mineral molybdenum. The molybdenum-sulfide (Mo-S) complex binds the substrates and transfers the electrons required for the oxidation reactions.

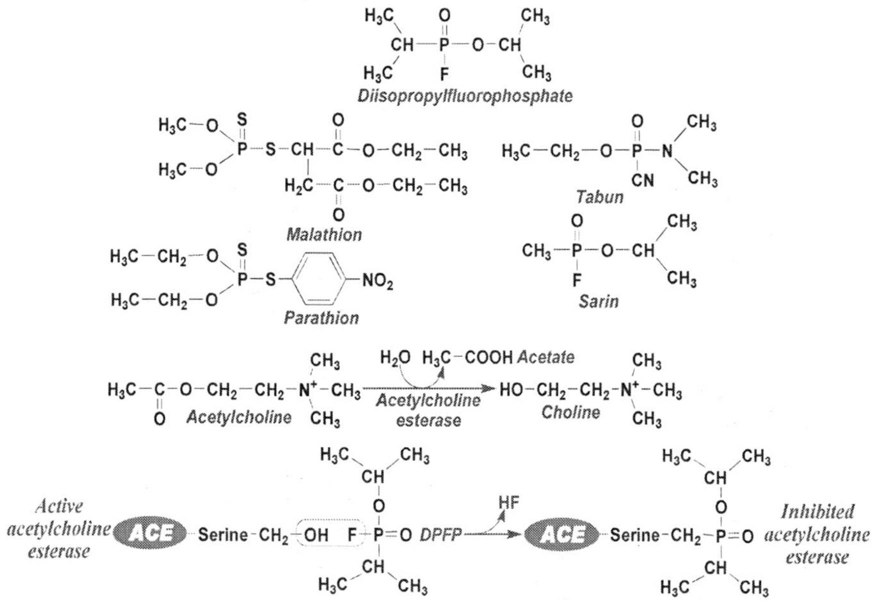

Figure 19. Organophosphorus compounds and the suicidal irreversible mechanism-based inhibition of the enzyme acetylcholine esterase by diisopropylfluorophosphate. Malathion and parathion are organophosphorus insecticides. The nerve gases Tabun and Sarin are other organophosphorus compounds.

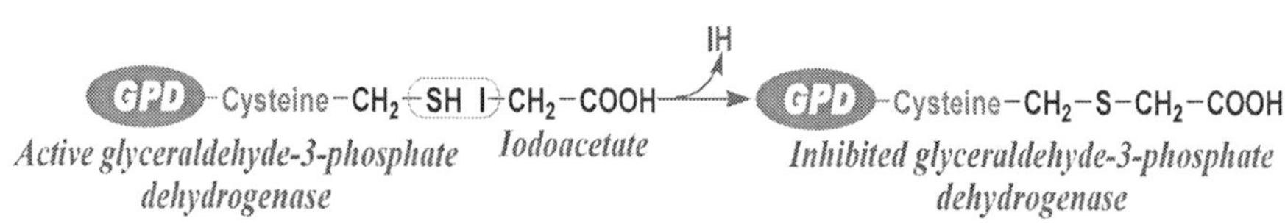

Figure 20. The suicidal irreversible mechanism-based inhibition of the enzyme glyceraldehyde-3-phosphate dehydrogenase by iodoacetate.

They also include the activated form of the guanosine analogue antiviral drug aciclovir - acycloguanosine (2-amino-9-((2-hydroxyethoxy)methyl)-1H-purin-6(9H)-one), as one of the most commonly-used antiviral drugs, it is primarily used for the treatment of herpes simplex and herpes zoster (shingles) viral infections. Aciclovir (see Figure 22) started a new era in antiviral therapy, as it is extremely selective and low in cytotoxicity. Aciclovir as a prodrug differs from previous nucleoside analogues in that it contains only a partial nucleoside structure: the sugar ring is replaced by an open-chain structure. It is selectively converted into acyclo-guanosine monophosphate (acyclo-GMP) by viral thymidine kinase, which is far more effective (3000 times) in phosphorylation than cellular thymidine kinase. Subsequently, the monophosphate form is further phosphorylated into the active triphosphate form, acyclo-guanosine triphosphate (acyclo-GTP), by cellular kinases. Acyclo-GTP is a very potent inhibitor of viral DNA polymerase; it has approximately 100 times greater affinity for viral than cellular polymerase. As a substrate, acyclo-GTP is incorporated into viral DNA, resulting in chain termination. Acyclo-GTP is fairly rapidly metabolized within the cell, possibly by cellular phosphatases.

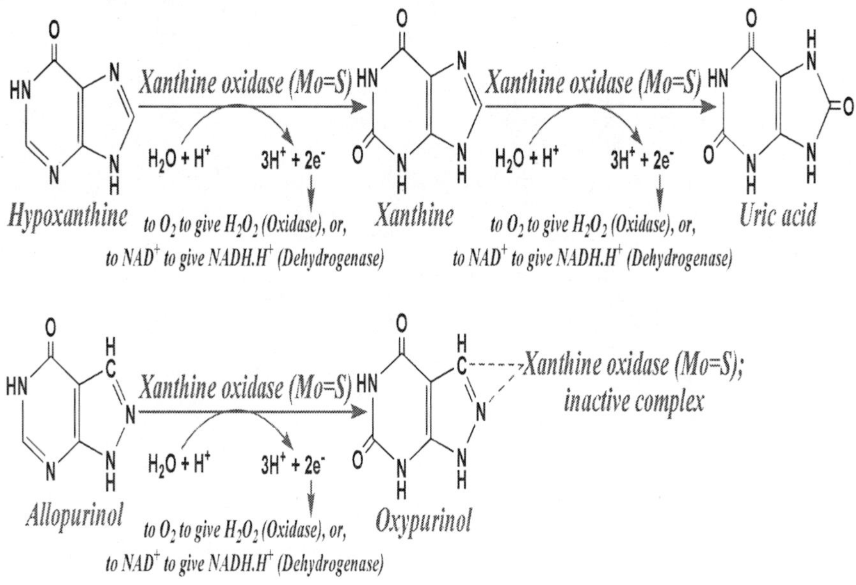

Figure 21. The suicidal irreversible mechanism-based inhibition of the enzyme xanthine oxidase by allopurinol.

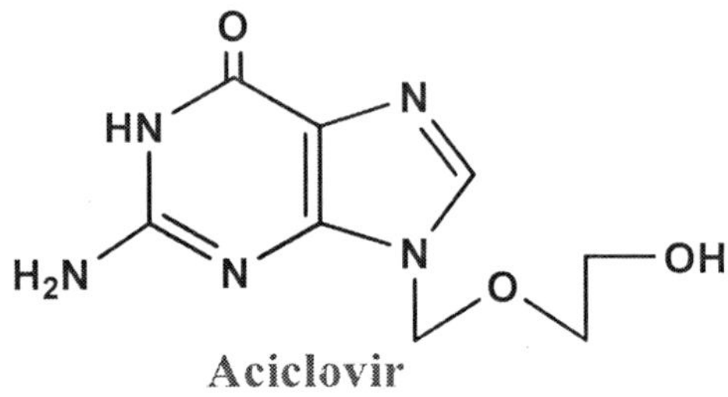

Figure 22. Aciclovir; the prodrug for the suicidal irreversible inhibition of the viral DNA polymerase.

The antibiotic penicillin is another transition state analog suicidal inhibitor that binds irreversibly covalently to serine at the active site of the bacterial enzyme glycopeptide transpeptidase. The enzyme is a serine protease required for synthesis of the bacterial cell wall and is essential for bacterial growth and survival. It normally cleaves the peptide bond between two D-alanine residues in a polypeptide. Penicillin structure contains a strained peptide bond within the β-lactam ring that resembles the transition state of the normal cleavage reaction, and thus penicillin binds very readily to the enzyme active site. The partial reaction to cleave the imitating penicillin peptide bond activates penicillin to bind irreversibly covalently to the active site serine (Figure 23).

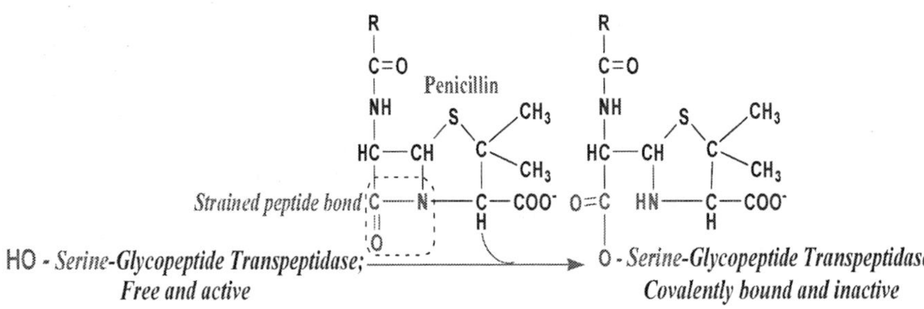

Figure 23. The suicidal irreversible mechanism-based inhibition of the bacterial enzyme glycopeptide transpeptidase by the antibiotic penicillin.

Aspirin (acetylsalicylic acid) provides an example of a pharmacologic drug that exerts its effect through the covalent acetylation of an active site serine in the enzyme cyclooxygenase (prostaglandin endoperoxide synthase). Aspirin resembles a portion of the prostaglandin precursor that is a physiologic substrate for the enzyme.

Heavy metal toxicity is caused by tight binding of a metal such as mercury, lead, aluminum, or iron, to a functional group at the active site of an enzyme. At high concentration of the toxin, heavy metals are relatively nonspecific for the enzymes they inhibit and inhibit a large number of enzymes. For example, it is impossible to specify which particular enzyme is implicated in mercury toxicity that binds reactive -SH groups at the active sites. Lead developmental and neurologic toxicity is caused by its ability to replace the normal functional metal in target enzymes; particularly Ca^{2+} in important enzymes, e.g., Ca^{2+}-calmodulin and protein kinase C. Because of their irreversible effect, heavy metals are routinely use as fixatives in histological preparations.

Kinetically, the irreversible inhibitors decrease the concentration of active enzyme and in turn decrease the maximum possible concentration of ES complex with ultimate reduction in the reaction rate of the inactivated individual enzyme molecules. The remaining unmodified enzyme molecules are normally functional considering their turnover number and K_m.

EXTRA ATTENTION: NATURAL POISONS AS ENZYME INHIBITORS AND INHIBITORY ENZYMES

Animals and plants have evolved to synthesize a vast array of poisonous products including secondary metabolites, peptides and proteins that can act as enzyme inhibitors. Natural toxins are usually small organic molecules and are so diverse that there are probably natural inhibitors for most metabolic processes. The metabolic processes targeted by natural poisons encompass more than enzymes in metabolic pathways and non-catalytic proteins. Many natural poisons act as neurotoxins. Some of these natural inhibitors, despite their toxic attributes, are valuable for therapeutic uses at lower doses. An example of a neurotoxin are the glycoalkaloids, from the plant species in the Solanaceae family (includes potato, tomato and eggplant), that are acetylcholinesterase inhibitors causing an uncontrolled increase in the acetylcholine neurotransmitter, muscular paralysis and then death. Although many natural toxins are secondary metabolites, these poisons also include peptides and proteins. An example of a toxic peptide is alpha-amanitin, which is found in relatives of the death cap mushroom. This is a potent enzyme inhibitor, in this case preventing the RNA polymerase II enzyme from transcribing DNA. The algal toxin microcystin is also a peptide and is an inhibitor of protein phosphatases. This toxin can contaminate water supplies after algal blooms and is a known carcinogen that can also cause acute liver hemorrhage and death at higher doses. Proteins can also be natural poisons or antinutrients, such as the trypsin inhibitors that are found in some legumes, potato, and tomato. Several invertebrate and vertebrate venoms contain protein and peptide enzyme inhibitors for, e.g., plasmin, renin and angiotensin converting enzymes. Inhibitory enzymes are enzymes that irreversibley inhibit other enzymes by chemically modifying them. In the broad sense, they include all proteases and lysosomal enzymes. Some of them are toxic plant products, e.g., ricin, a glycosidase that is an extremely potent protein toxin found in castor oil beans. It inactivates ribosomes by cleavage the eukaryotic 28S rRNA and reduces protein synthesis and a single molecule of ricin is enough to kill a cell.

Reversible Inhibition

They may be competitive, noncompetitive, or uncompetitive inhibitors relative to a particular substrate. Products of enzymatic reactions are reversible inhibitors of the producing enzymes. A decrease in the rate of an enzyme caused by the accumulation of its own product plays an important role in the balance and most economic usage of metabolic pathways. This prevents one enzyme in a sequence of reactions from generating a product more than the capacity of the next enzyme in that sequence, e.g., inhibition of hexokinase by accumulating glucose 6-phosphate.

With the reduction in the inhibitor concentration, the enzyme activity is regenerated due to the non-covalent association and the reversible equilibrium with the enzyme. The equilibrium constant for the dissociation of enzyme inhibitor complexes is known as K_i that equals [E][I]/[EI]. The efffect of K_i on the reaction kinetics is reflected on the normal K_m and or V_{max} observed in Lineweaver-Burk plots; in a pattern dependent on the type of the inhibitor. The inhibitor is removable by several ways. *The three common types of reversible inhibitions are:*

- Competitive reversible inhibition.
- Uncompetitive reversible inhibition.
- Mixed reversible inhibition (or non-competitive inhibition).

COMPETITIVE REVERSIBLE INHIBITION

The competitive inhibitor is structurally related to the substrate and binds reversibly at the active site of enzyme and occupies it in a mutually exclusive manner with the substrate. Therefore, the competitive inhibitor competes with the substrate for the active site. The binding is mutually exclusive because of their free competition. According to the law of mass action, relatively higher inhibitor concentration prevents the substrate binding. Since the reaction rate is directly proportional to [ES], reduction in ES formation for EI formation lowers the rate. Increasing substrate towards a saturating concentration alleviates competitive inhibition. In the time enzyme-substrate complex releases the free enzyme and a product, the enzyme-inhibitor complex does release neither free enzyme nor a product. Reversible inhibition is of short duration in the biological system because it depends on substrate availability and/or rate of the catabolic clearance of the inhibitor (Figure 24).

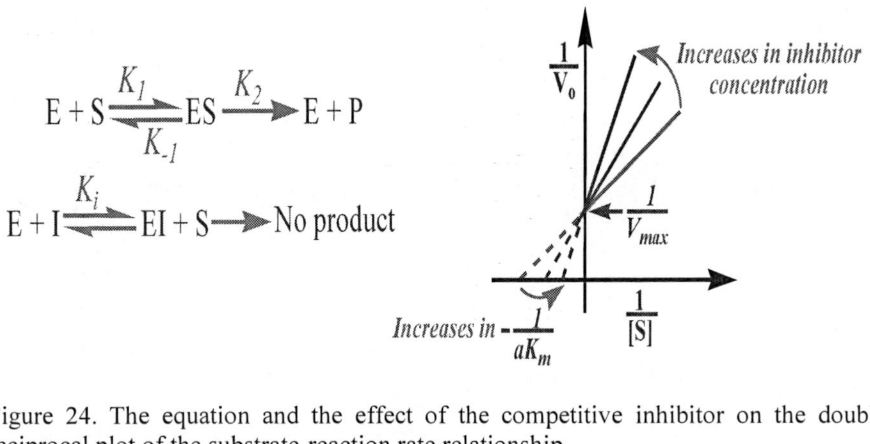

Figure 24. The equation and the effect of the competitive inhibitor on the double reciprocal plot of the substrate-reaction rate relationship.

Kinetically, the inhibitor (I) binds the free enzyme reversibly to form enzyme inhibitor complex (EI) that is catalytically inactive and cannot bind the substrate. The competitive inhibitor reduces the availability of free enzyme for the substrate binding. Thus, the K_m of the normal reaction is increased to a new K_m (aK_m) as a function of the inhibitor concentration (expressed in the "a" factor - *a*pparent K_m in presence of the inhibitors), where the substrate concentration at $V_o = \frac{1}{2} V_{max}$ is equal to aK_m. The "a" can be calculated from the change in the slope of the line at a given inhibitor concentration;

$$a = 1 + \frac{[I]}{K_I},\ where,\ K_I = \frac{[E][I]}{[EI]}$$

Therefore, competitive inhibitors do not affect the turnover number (active site catalysis per unit time) or the efficiency of the enzyme because once free the enzyme behaves normally. The Michaelis-Menten equation for competitive inhibitors becomes

$$V_o = \frac{V_{max}[S]}{aK_m + [S]}$$

Consequently, the double reciprocal form of the equation is also modified so as the line slope becomes

$$\frac{aK_m}{V_{max}}$$

And the intercept with y-Axis stays at

$$\frac{1}{V_{max}}$$

But the intercept with the x-axis at

$$-\frac{1}{aK_m}$$

will differ according to the concentration of the competitive inhibitor. The later property is characteristic for competitive inhibitors.

Examples include the classical competitive inhibitory effect of *malonic acid* on succinate dehydrogenase (SD) of the Krebs' cycle that reversibly dehydrogenates succinate into fumarate. Other less potent competitive inhibitors of succinate dehydrogenase include; oxalate, glutamate and oxaloacetate. The common molecular geometric feature of these compounds is the presence of two negatively charged -COOH groups suggesting that the active site of the flavoprotein SD has specifically positioned two positively charged binding groups (Figure 25).

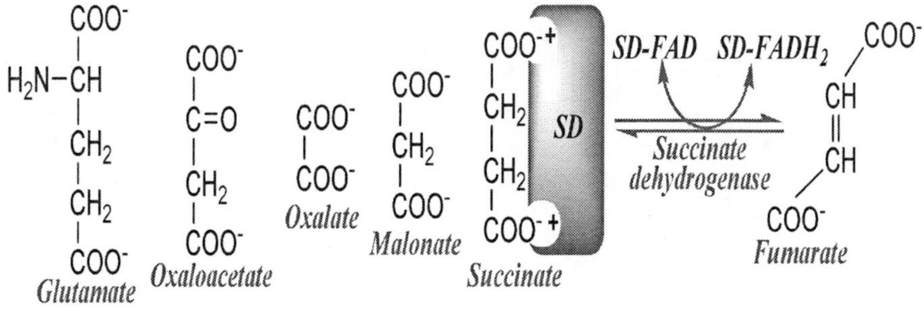

Figure 25. The substrate and different competitive inhibitors of succinate dehydrogenase (SD).

Methotrexate - competitive inhibitor of dihydrofolate reductase (DHFR) is another example. The drug is used as anticancer antimetabolite chemotherapy particularly for pediatric leukemia. It hinders the availability of tetrahydrofolate as a carrier for one-carbon moieties important for anabolic pathways -particularly synthesis of purine nucleotides for DNA replication (Figure 26).

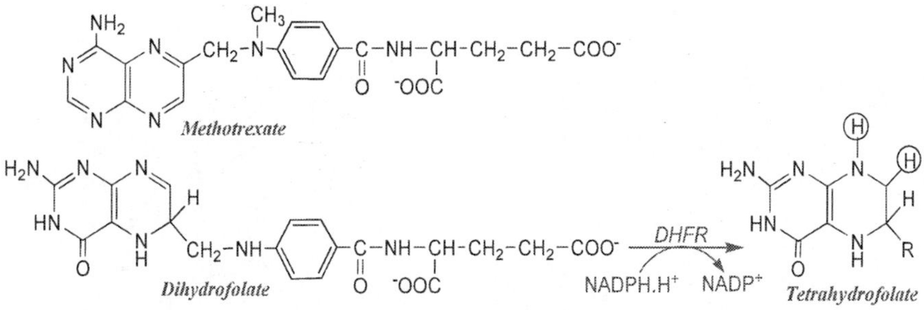

Figure 26. The substrate and methotrexate as a competitive inhibitor for dihydrofolate reductase.

Sulfanilamides - the simplest form of Sulfa drugs - were among earliest antibacterial chemotherapeutic drugs classified as enzyme inhibitors. They are competitive inhibitors of the bacterial folic acid synthesizing enzyme system from *p*-aminobenzoic acid. Bacterial cannot absorb pre-made folate that is necessary to be synthesized *de novo*. Structural similarity of sulfanilamide (and other sulfas derived from it) to *p*-aminobenzoic acid made them competitive inhibitors to the enzyme (Figure 27).

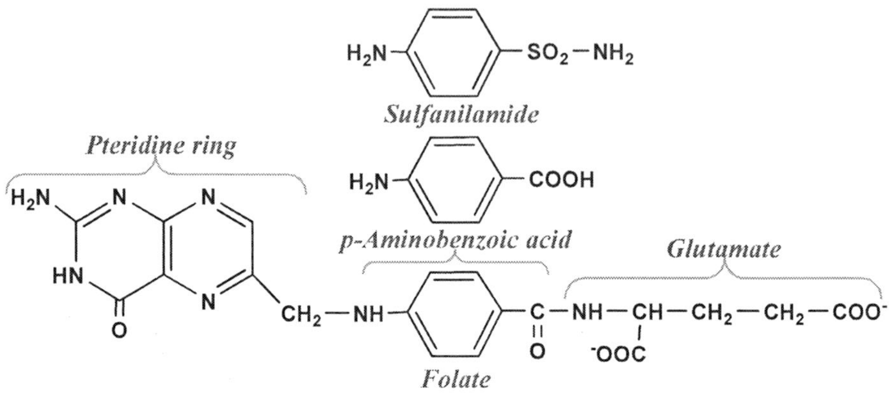

Figure 27. The *p*-aminobenzoic acid substrate and sulfanilamide as a competitive inhibitor during the bacterial folate synthesis.

Male erectile impotence was a major medical problem till the recent discovery of a group of chemicals with molecular structural similarity to cGMP that competitively inhibit the cGMP-phosphodiesterase-5. They include sildenafil citrate (Viagra; Figure 28), vardenafil (Levitra) and tadalafil (Cialis). The inhibition of this enzyme that has a limited tissue distribution including the penile cavernous tissue spares cGMP. Accumulation of cGMP leads to smooth muscle relaxation (vasodilation) of the intimal cushions of the helicine arteries, resulting in increased inflow of blood and an erection.

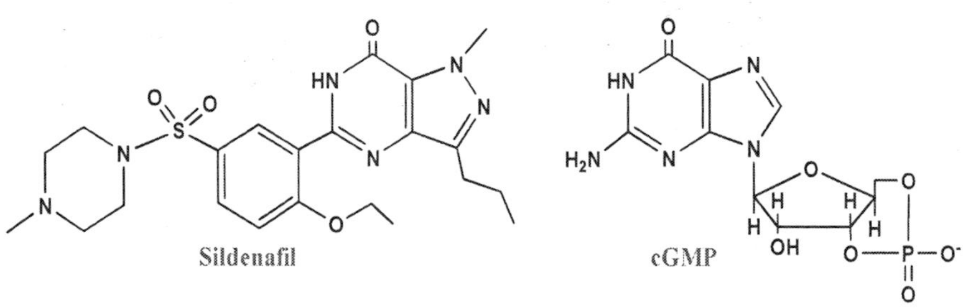

Figure 28. The cGMP substrate and sildenafil a competitive inhibitor of the cGMP-phosphodiesterase-5.

Another example of these substrate mimics competitive inhibitors are the peptide-based protease inhibitors, a very successful class of antiretroviral drugs used to treat HIV, e.g., ritonavir that contains three peptide bonds (see Figure 29).

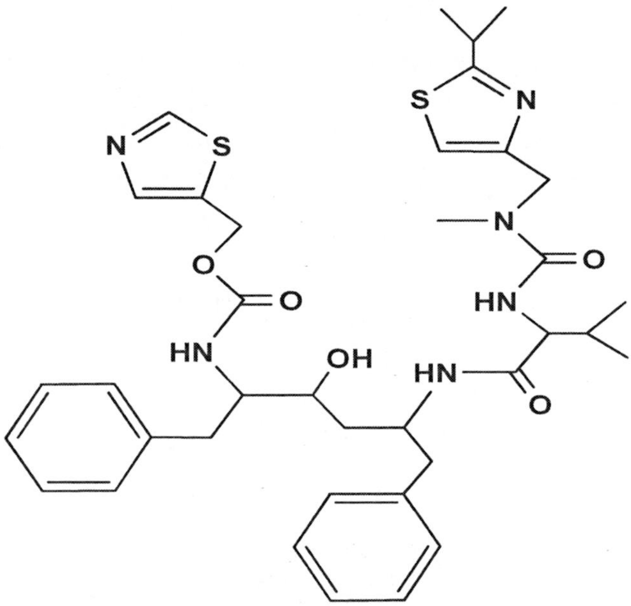

Figure 29. The peptide-based competitive protease inhibitor ritonavir.

Reversible competitive inhibitors of acetylcholinesterase, such as edrophonium, physostigmine, and neostigmine, are used in the treatment of myasthenia gravis and in anesthesia. The carbamate pesticides are also examples of reversible acetylcholinesterase inhibitors.

UNCOMPETITIVE REVERSIBLE INHIBITION

Uncompetitive inhibitor has no structural similarity to the substrate and does not bind the free enzyme but binds the enzyme after complexation with the substrate that exposes the inhibitor binding site (ESI). Its binding, although away from the active site, causes structural distortion of the active and allosteric sites of the complexed enzyme that inactives catalysis. This leads to a decrease in both K_m and V_{max}. Increasing substrate towards a saturating concentration does not reverse this type of inhibition and reversal requires special treatment, e.g., dialysis. This type of inhibition is also encountered in multi-substrate enzymes, where the inhibitor competes with one substrate (S2) to which it has some structural similarity and is uncompetitive for the other (S1). The reaction without the inhibitor would be; $E + S_1 \Leftrightarrow ES_1 + S2 \Leftrightarrow$

$ES_1S_2 \Rightarrow E + Ps$ and with uncompetitive inhibitor becomes; $E + S_1 \Leftrightarrow ES_1 + I \Rightarrow ES_1I$ (prevents S2 binding) $\Rightarrow$ no product. It is a rare type and the inhibitor may be the reaction product or a product analog.

Kinetically, uncompetitive inhibition modifies the Michaelis-Menten equation by (a') factor that proportionates with the inhibitor concentration to be,

$$V_o = \frac{V_{max}[S]}{K_m + a'[S]}$$

and in the double-reciprocal equation to be,

$$\frac{1}{V_o} = \frac{a'}{V_{max}} + \frac{K_m}{V_{max}} X \frac{1}{[S]}$$

and y-intercept is at

$$\frac{a'}{V_{max}}$$

while x-intercept is at

$$-\frac{a'}{K_m}$$

whereas, the line slope stays

$$\frac{K_m}{V_{max}}$$

This gives a number of lines in the Lineweaver-Burk plot that are parallel to the normal line with decreased $1/V_{max}$ and $-a'/K_m$ proportional to concentrations of the uncompetitive inhibitor. The later is characteristic to uncompetitive inhibition (Figure 30).

This type of inhibition is rare, but may occur in multimeric enzymes. Examples of uncompetitive reversible inhibitors include; inhibition of lactate dehydrogenase by oxalate; inhibition of alkaline phosphatase (EC 3.1.3.1) by L-phenylalanine, and, inhibition of the key regulatory heme synthetic enzyme; δ-aminolevulinate synthase and dehydratase and heme synthetase by heavy metal ion, e.g., lead. Heavy metals, e.g., lead, form mercaptides with -SH at the active site of the enzyme (2 R-SH + Pb $\Rightarrow$ R-S-Pb-S-R + 2H). Oxidizing agents, e.g., ferricyanide also oxidizes -SH into a disulfide linkage (2 R-SH ⇒ R-S-S-R). Reversion here requires treatment with reducing agents and/or dialysis.

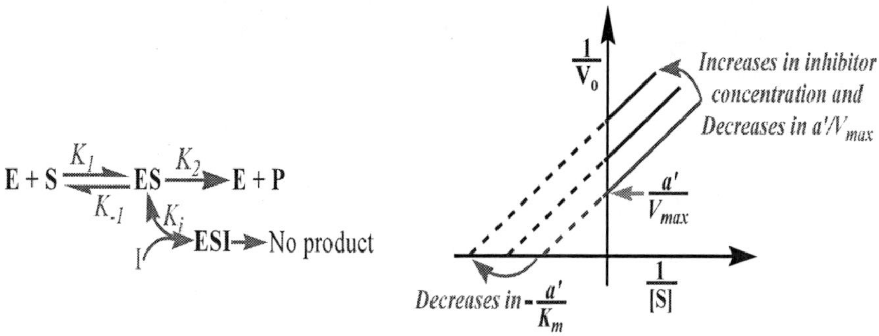

Figure 30. The equation and the effect of the uncompetitive inhibitor on the double reciprocal plot of the substrate-reaction rate relationship.

Mixed (Noncompetitive) Inhibition

The mixed type inhibitor does not have structural similarity to the substrate but it binds both of the free enzyme and the enzyme-substrate complex. Thus, its binding manner is not mutually exclusive with the substrate and the presence of a substrate has no influence on the ability of a non-competitive inhibitor to bind an enzyme and *vice versa*. However, its binding - although away from the active site - alters the conformation of the enzyme and reduces its catalytic activity due to changes in the nature of the catalytic groups at the active site. EI and ESI complexes are nonproductive and increasing substrate to a saturating concentration does not reverse the inhibition leading to unaltered K_m but reduced V_{max}. Reversal of the inhibition requires a special treatment, e.g., dialysis or pH adjustment. Some classifications differentiate between non-competitive inhibition as defined above and *mixed inhibition* in that the EIS-complex has residual enzymatic activity in the mixed inhibition.

Kinetically, mixed type inhibition causes changes in the Michaelis-Menten equation so as

$$V_o = \frac{V_{max}[S]}{aK_m + a'[S]}$$

Thus, mixed type inhibition - as the name imply - has a change in the denominator with K_m modified by factor (a) as in competitive inhibition, and [S] modified by factor (a') as in uncompetitive inhibition. In the double reciprocal equation,

$$\frac{1}{V_o} = \frac{a'}{V_{max}} + \frac{aK_m}{V_{max}} X \frac{1}{[S]}$$

the line slope is

$$\frac{aK_m}{V_{max}}$$

and the intercept with y-axis is at

$$\frac{a'}{V_{max}}$$

and with x-axis is at

$$\frac{a'}{aK_m}$$

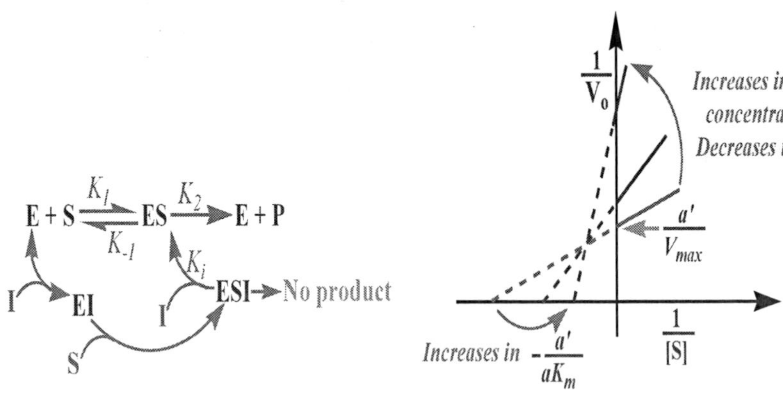

Figure 31. The equation and the effect of the mixed type (noncompetitive) inhibitor on the double reciprocal plot of substrate-reaction rate relationship.

This results in progressive decreases in V_{max} and progressive increases in K_m proportional to the increase in the mixed inhibitor concentration. The double reciprocal plot shows a number of lines reflecting decreases in V_{max}/increases in K_m but their intercept is to the left of the y-axis. Mixed type inhibitor would be called non-competitive only if [a = a'], where, it will only lower V_{max} without affecting the K_m (Figure 31).

Table 3. Comparison of the different types of reversible inhibition.

Type	Nature of the inhibitor binding	Fate of the inhibition
Competitive Inhibitor	The inhibitor binds the catalytic/substrate binding site. It competes with substrate for binding. Inhibition is reversible by increasing substrate concentration.	So as not to decrease V_{max}, the substrate concentration has to be increased as reflected on increased K_m.
Uncompetitive Inhibitor	Substrate binding exposes the inhibitor binding site away from the catalytic/substrate binding site. Increasing substrate concentration does not reverse the inhibition.	The inhibited reaction rate parallel the normal one as reflected on decreased both V_{max} and K_m.
Mixed (noncompetitive) Inhibitor	The inhibitor binds each of the free enzyme and the substrate-enzyme complex away from the catalytic/substrate binding site. Increasing substrate concentration does not reverse the inhibition.	Only the V_{max} is decreased proportionately to inhibitor concentration, whereas, K_m is unchanged since increasing substrate concentration is ineffective.

Examples of noncompetitive inhibitors are mostly poisons because of the crucial role of the targeted enzymes. Cyanide and azide inhibits enzymes with iron or copper as a component of the active site or the prosthetic group, e.g., cytochrome c oxidase (EC 1.9.3.1). They include the inhibition of an enzyme by hydrogen ion at the acidic side and by the hydroxyl ion at the alkaline side of its optimum pH. They also include inhibition of; carbonic anhydrase by acetazolamide; cyclooxygenase by aspirin; and, fructose-1,6-diphosphatase by AMP. Cyanide binds to the Fe^{3+} in the heme of the cytochrome aa_3 component of cytochrome c oxidase and prevents electron transport to O_2. Mitochondrial respiration and energy production cease, and cell death rapidly occurs. The central nervous system is the primary target for cyanide toxicity. Acute inhalation of high concentrations of cyanide (e.g., smoke inhalation during a fire and automobile exhaust) provokes a brief central nervous system stimulation rapidly followed by convulsion, coma, and death. Acute exposure to lower amounts can cause lightheadedness, breathlessness, dizziness, numbness, and headaches. Cyanide is present in the air as hydrogen cyanide (HCN), in soil and water as cyanide salts (e.g., NaCN), and in foods as cyanoglycosides. Comparison of the three types of the reversible enzyme inhibitors is presented in Table 3.

In a special case, the mechanism of *partially competitive inhibition* is similar to that of non-competitive, except that the EIS complex has catalytic activity, which may be lower or even higher (partially competitive activation) than that of the enzyme-substrate (ES) complex. This inhibition typically displays a lower V_{max}, but an unaffected K_m value.

Extra Attention

- ***Slow-tight inhibition:*** Slow-tight inhibition occurs when the initial enzyme-inhibitor complex EI undergoes isomerizing conformational change to a more tightly binding complex. However, the overall inhibition process is reversible. This manifests itself as slowly increasing enzyme inhibition. Under these conditions, traditional Michaelis-Menten kinetics gives a false value of a time-dependent K_i. The true value of K_i can be obtained through more complex analysis of the on (k_{on}) and off (k_{off}) rate constants for inhibitor association.
- ***Substrate and product inhibition:*** Substrate and product inhibition is where either the substrate or product of an enzyme reaction inhibits the enzyme's activity. This inhibition may follow the competitive, uncompetitive or mixed patterns. In substrate inhibition there is a progressive decrease in activity at high substrate concentrations. This may indicate the existence of two substrate-binding sites in the enzyme. At low substrate, the high-affinity site is occupied and normal kinetics is followed. However, at higher concentrations, the

second inhibitory site becomes occupied, inhibiting the enzyme. Product inhibition is often a regulatory feature in metabolism and can also be a form of negative feedback; see allosteric regulation.

- ***Antimetabolites:*** They are chemicals that interfere with the normal metabolism of normal biochemical metabolite(s). This in most of case is due to their structural similarity to such physiological substrates and therefore works as competitive enzyme inhibitors. They include antifolates such as methotrexate, hydroxyurea and purine and pyrimidine analogues. They are mainly used as cytotoxic anticancer drugs through inhibiting DNA and RNA synthesis and cell division.

ANTIENZYME

Intestinal parasites, e.g., Ascaris, protect themselves from digestion by expressing on their surface substances that are protein in nature which inhibit the action of digestive enzymes, e.g., pepsin and trypsin. The blood plasma and extracellular fluids are containing several types of protease inhibitors particularly important in controlling the blood clot formation and dissolution and matrix and cytokine homeostasis. Most of these inhibitors are peptides and several of them are also isolated from raw egg white, potatoes, tomatoes and Soya bean and other plant sources. Most of the natural peptide protease inhibitors are similar in structure to the amino acid sequence of the peptide substrates of the enzyme. Designed peptide protease inhibitors are important drugs, e.g., captopril that is a metalloprotease angiotensin-converting enzyme peptide inhibitor. Inhibiting this enzyme prevent activation of angiotensin and therefore prevent vasoconstriction to lower blood pressure. Crixivan is an anti-retroviral aspartyl protease peptide inhibitor used in the treatment of Human Immunodeficiency Virus (HIV)-induce acquired immunodeficiency syndrome (AIDS). It inhibits the HIV protease that cleaves the large multidomain viral protein into active enzyme subunits. Because these peptide inhibitors may not be specific, they have several side-effects as drugs.

Antibodies against several nonfunctional plasma enzymes have clinical diagnostic importance since they are longer living than the enzyme itself and hence reflect the disease history better. In this respect, autoimmune antibodies are clinically important in diagnosis of autoimmune diseases, e.g., anti-glutamic acid decarboxylase antibodies in type 1 diabetes mellitus.

Extra Attention: Effect of Radiations, Light and Oxidants on the Rate of the Enzyme Catalyzed Reaction

Light inhibits most enzyme activity although some enzymes, e.g., amylase are activated by red or green light and also specific DNA repairing enzymes (e.g., UV-specific endonuclease) are activated by the blue and UV light. Ultraviolet rays and ionizing radiations cause denaturation of most enzymes. Most enzymes contain sulfhydryl (-SH) groups at their active sites which upon oxidation by oxidants and free radicals by oxidants and free radicals inactivate the enzyme.

Chapter VII

REGULATION OF ENZYME ACTIVITY

INTRODUCTION

Other than the aforementioned inherent kinetic prosperities of enzymes and the enzyme interaction with factors modulating their activity (including inhibitors) as micro-controlling mechanisms, the global regulation of the complex network of intracellular and extracellular enzymatic reactions to the maximum economy of the biological system is executed by:

- Allosteric regulation.
- Compartmentation of enzymes.
- Hormonal control and covalent modification.
- Regulation of enzyme half-life (rate of synthesis vs. rate of degradation).
- Synthesis of enzymes in a proenzyme (zymogen) form and control of their activation.
- Expression of different tissue- or cell compartments-specific isoenzyme forms.

ALLOSTERIC (FEEDBACK) REGULATION

Other than simple enzymes regulated by interaction with substrates and/or inhibitors, there is another class of enzymes that are also regulated through binding into low molecular weight physiologically important molecules called allosteric effectors (activator/inhibitors). These molecules modulate the activity of such enzymes in ways different from those induced by substrate and/or non-allosteric inhibitor through binding at an allosteric site (allo = the other, and steric = steering). Thus, allosteric enzymes are those key regulatory enzymes susceptible for regulation by allosteric effectors. Most allosteric enzymes are multimeric proteins composed of two or more polypeptide subunits; with two or more catalytic subunits. Examples are acetyl-CoA carboxylase that is monomeric when is inactive and polymerizes when is activated, and isocitrate dehydrogenase with 4 subunits per molecule. *Subunits in each enzyme may attain one of two conformational states;* i) with very low enzymatic activity, or, ii) with very high enzymatic activity.

The effector may bind directly to its specific binding sites in the enzyme, or indirect, where the effector binds to other regulatory proteins or protein subunits that interact with or dissociates from the allosteric enzyme and thus influence catalytic activity. The allosteric effector may have a high, little or no structural similarity to the substrates or coenzyme. They bind non-covalently at the allosteric site and alter the conformation of the enzyme. This change modulates substrate binding affinity (K_m) or the catalytic efficiency (V_{max}) of the enzyme. Example is the allosteric activation of hexokinase by AMP and Pi, and, its allosteric inhibition by glucose-6-phosphate.

The kinetics of allosteric enzyme reaction has a sigmoid saturation curve and does not follow the hyperbolic Michaelis-Menten V_0/[S] relationship. Therefore, 1/2 V_{max} value does not correlate [S] corresponding to K_m. Instead, the symbol $K_{0.5}$ is used to represent [S] giving 1/2 V_{max}. Sigmoid kinetic reflects cooperative interactions between protein subunits mediated by non-covalent bonds that are modulable with the allosteric effector.

When the substrate at the same active site of the multi-subunit enzyme works as substrate and allosteric regulator, the process is called *homotropic allosteric regulation* (e.g., allosteric activation of glycogen synthase with glucose-6-phosphate, and, pyruvate dehydrogenase with pyruvate). It can act as such, either by binding to the substrate-binding site, or to an allosteric effector site. The kinetics curve resembles that of hemoglobin-O_2 saturation curve. The subunits act cooperatively: the binding of one molecule of substrate to one binding site alters the enzyme's conformation and enhances the binding of subsequent substrate molecules. This accounts for the sigmoid rather than hyperbolic change in V_0 with increasing [S], where, a small change in the concentration of a modulator causing a large change in the enzyme activity.

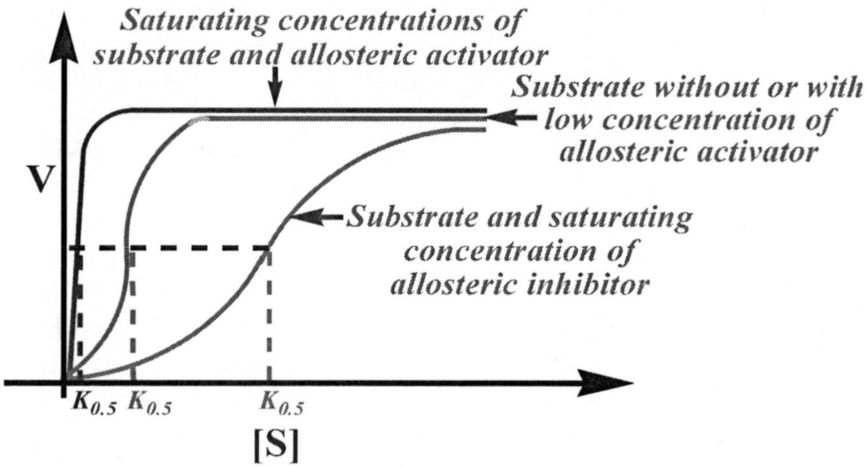

Figure 32. The effect of the allosteric effectors (activator/inhibitor) on the substrate-reaction rate relationship.

In the *heterotropic allosteric regulation*, the modulator(s) is a metabolite in the reactions of a cascade or a related reaction pathway (e.g., allosteric activation of isocitrate dehydrogenase by ADP). A heterotropic activator may cause the curve to become more nearly hyperbolic, increased reaction velocity with a decrease in $K_{0.5}$ but no change in V_{max} at a fixed substrate concentration; or, it may increase V_{max} with little change in $K_{0.5}$. A negative modulator may produce a *more* sigmoid substrate-saturation curve, with an increase in $K_{0.5}$ or a lower V_{max}. *Enzymes with heterotropic allosteric regulation display two types of velocity versus substrate concentration*; i) *hyperbola curve* when they are fully activated by saturating concentration of an activator and substrate, and, ii) *sigmoid curve* that has two states; a) without or with very low activator concentration where the enzyme will be sensitive to fluctuation in substrate concentration and attains a near hyperbolic curve, or, b) presence of inhibitor and low or intermediate substrate concentration, where, the enzyme attains a sigmoid low activity curve (Figure 32). The three conditions do not alter V_{max} but the $K_{0.5}$ changes significantly. Because of that, this is called K-type allosteric regulation (enzyme and effector). A very rare type of allosteric enzymes shows a change in V_{max} (increase/decrease) with nearly constant $K_{0.5}$. Because of that, this is called V-type allosteric regulation (enzyme and effector).

Allosteric enzymes could attain a lower molecular weight upon the activation-induced conformational change, e.g., the protein kinase A that has 4 subunits (2 regulatory and two catalytic) and upon activation through binding to cAMP it loses the two regulatory subunits and each of the catalytic subunits will be independently catalytically active. Another example of such type of allosteric enzymes is the aspartate transcarbamoylase that become smaller after the transfer of the carbamoyl moiety into aspartate and their releases as carbamoyl aspartate. Other allosteric enzymes stay with constant molecular weight, e.g., phosphofructokinase-1 (EC 2.7.1.11) and isocitrate dehydrogenase. Still other allosteric enzymes gain higher molecular weight, e.g., acetyl-CoA carboxylase due to its polymerization upon activation.

The binding of the allosteric effector to the different subunit of an allosteric enzyme that explain the sigmoid nature of the curve of their kinetic could be sequential (after Koshland et al) or concerted (after Monod et al). In the sequential model, the ligand binding into one subunit causes conformational change that is transmitted to a second subunit then to a third and so on. Consequently an enzyme exists in a spectrum of conformational states, i.e., completely inactive, intermediate hybrid states, or completely active. In the concerted symmetric model, there are no intermediate states and the enzyme is either active or inactive, i.e., all subunits get activated upon binding.

Therefore, allosteric effectors could be positive (i.e., stimulatory by increasing enzyme-substrate affinity that lowers K_m or increases V_{max}) or negative allosteric effectors (i.e., inhibitory by decreasing enzyme-substrate affinity that increases K_m or lowers V_{max}). In a metabolic pathway (i.e., an ordered linear or circular cascade of reactions, e.g., glycolysis or the citric acid cycle), a succeeding enzyme uses the product of preceding enzyme as a substrate etc. The end product(s) of such pathway are often non-competitive inhibitors for one of the enzymes catalyzing initial steps of the pathway (usually the first irreversible step, i.e., *the committed step*), thus regulating the amount of end product made by the pathway. This is *the classical basic feedback allosteric inhibition mechanism.*

Generally, feedback inhibition refers to the phenomenon whereby an immediate or a late product in a cascade (or related cascades) of catabolic or anabolic reactions allosterically inhibits a key regulatory enzyme active at the early steps of the pathway. For this reason rapid disposal of end products is essential for increasing the rate and continuation of the pathway. Moreover, this mechanism is important for the economic usage of these pathways where further production is unnecessary if the product is accumulating. Aspartate transcarbamoylase is the second step and a key regulatory allosteric enzyme during pyrimidine biosynthesis where, CTP is an end-product. The later is a strong allosteric inhibitor for the transcarbamoylase. Major control of energy producing pathways is through feedback regulatory effect of allosteric markers of energy surplus (ATP, GTP, $NADH.H^+$, acetyl-CoA, citrate, pyruvate, glucose-6-phosphate and shift towards acidic pH) vs. energy shortage markers (Pi, ADP, AMP, NAD^+ and shift towards alkaline pH). For example, ATP and citrate are allosteric inhibitors, whereas, Fructose-2,6-diphosphate is allosteric activator for phosphofructokinase-1. Glucose-6-phosphate is allosteric inhibitor for hexokinase, whereas, it is an allosteric activator for glycogen synthase.

Other than the Basic Classic Type of Allosteric Feedback Inhibition, there are 4 Types of Feedback Allosteric Inhibition (Figure 33)

I. The basic feedback inhibition mechanism, where one abundant end product (P) inhibits one committed step.

II. *Sequential feedback inhibition,* where each abundant end product (P_1 or P_2) inhibits the upstream branch committed step. The abundance of both blocks the utilization of C leading into its abundance and in turn inhibition of the first common committed step of the whole pathway.

III. *Enzyme multiplicity,* where each abundant end product inhibits both the upstream branch committed step and one of the enzymes performing the first common committed step. Thus, a single allosteric end product inhibits several enzymes with different catalytic actions.

IV. *Concerted feedback inhibition,* where each abundant end product inhibits the upstream branch committed step, and all together, they inhibit the first common committed step. No single end product alone can inhibit the first common committed enzyme and when 2 or more allosteric end products exist simultaneously in excess an additive inhibition occurs. If such concerted inhibition is more than additive, the mechanism of allosteric inhibition is called *cooperative*.

Cumulative feedback inhibition, where each abundant end product inhibits the upstream branch committed step and each end product partially inhibits the first common committed step. Therefore, when two or more allosteric end products are in effect, the inhibition of the first common committed is strictly additive.

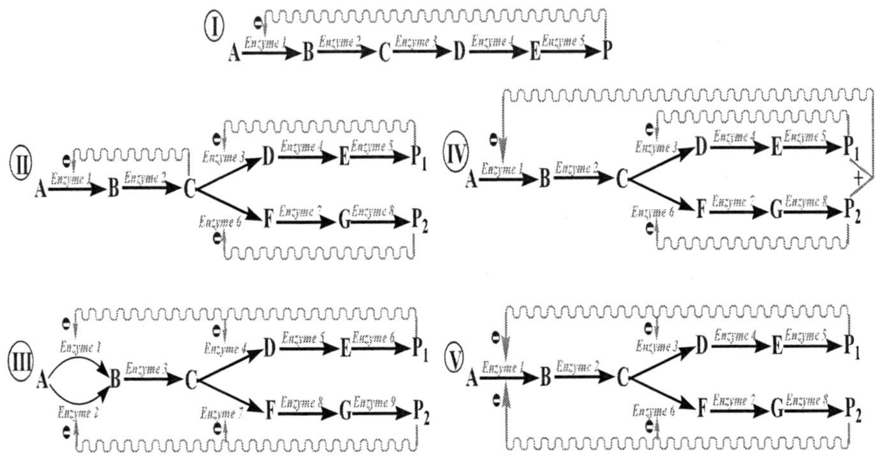

Figure 33. Common mechanisms of allosteric feedback inhibition. I is the basic feedback inhibition mechanism; II is the sequential feedback inhibition; III is enzyme multiplicity mode; IV is the concerted feedback inhibition; V is cumulative feedback inhibition.

COMPARTMENTATION OF THE ENZYME

This process allows creating an isolated space where specific coordinated functions are carried out. The outcome of metabolism depends both on availability of precursors and the compartment of the reaction. Thus, a mitochondrial enzyme would work only in the mitochondria because the appropriate conditions (substrate, withdrawal of the product and controlling elements) are only available there independent of metabolic pathways proceeding elsewhere. For example, although acetyl-CoA is available in the cytosol and mitochondria, it is used in the two compartments independently through regulated distinct pathways; fatty acid synthesis in the cytosol and citrate synthesis by Krebs' cycle in the mitochondria. Also, CO_2 is used in cytoplasm for pyrimidine synthesis, whereas, it is used in mitochondria for urea synthesis. However, some enzymes work in several compartments *albeit* differentially and not integrated in a similar cascade of reactions. For example, the isocitrate dehydrogenase is NAD/NADP-dependent in the mitochondria as an integral part of the citric acid cycle, whereas, the cytosolar form of the enzyme is NADP-dependent and works as a major source of NADPH.

HORMONAL CONTROL AND COVALENT MODIFICATION

Hormones such as insulin and glucocorticoids control enzyme activities through a slow pathway that regulates the rate of the gene expression (transcription-translation; See later) and rate of enzyme degradation. Hormones also have a rapid mechanism of controlling enzyme activities through a rapid cell membrane receptor-mediated covalent modification of individual enzyme molecules. Covalent modification is the addition or removal of a modifying chemical moiety to bind covalently to the enzyme molecule, e.g., phosphate, glucose, methyl, ADP-ribose, or acetyl groups, etc. This covalent attachment of a chemical moiety to the enzyme causes an activating/inactivating conformational change in the enzyme structure depending on its nature. Phosphorylation (mainly on -OH group of a serine/threonine but also a tyrosine residue) inhibits some enzymes such as glycogen synthase ("a"-active form becomes "b"-inactive form) while activating others, e.g., glycogen phosphorylase ("b"-inactive form becomes "a"-active form). Phosphorylation is catalyzed by protein kinases, whereas, dephosphorylation is catalyzed by protein phosphatases both are hormone-regulated. Hormones execute these effects through controlling the availability of kinase/phosphatase allosteric effector(s), e.g., cAMP, cGMP, inositol triphosphate, diacylglycerol and Ca^{2+}. Whereas antiinsulins, e.g., glucagon and adrenaline increase these effectors through activating their synthesis, insulin, on the other hand, lowers availability of cAMP by stimulating its degradation through its phosphodiesterase. There could be several covalent modification sites on one enzyme that are targeted by several regulatory mechanisms.

For example, metabolic antiinsulin hormones act mostly through activating adenylate cyclase that converts ATP into cAMP. The later binds the regulatory subunits of cAMP-dependent protein kinase enzyme to release each of its catalytic subunits free and active to phosphorylate substrate enzymes (Figure 34). Although the enzymatic covalent modification mechanisms are largely reversible, some are irreversible, e.g., the non-physiological ADP-ribosylation of the α-subunit of the stimulatory G-protein (Gs) of the intestinal mucosal by cholera toxin. This blocks the GTPase activity of this subunit and leaves the G-protein permanently active (αG_s-GTP complex) to activate adenylate cyclase that produces more cAMP. Accumulating cAMP activates the intestinal mucosal Cl^--channel to secrete Cl^- into the intestinal lumen accompanied with Na^+ and water causing the severe characteristic cholera diarrhea and dehydration. However, physiological poly ADP-ribosylation is rapidly reversed by poly(ADP-ribose) glycohyrolase. Other than the metabolism, covalent modification mechanisms also control a number of biological processes including; DNA organization and gene expression, cell proliferation and differentiation, and protein degradation.

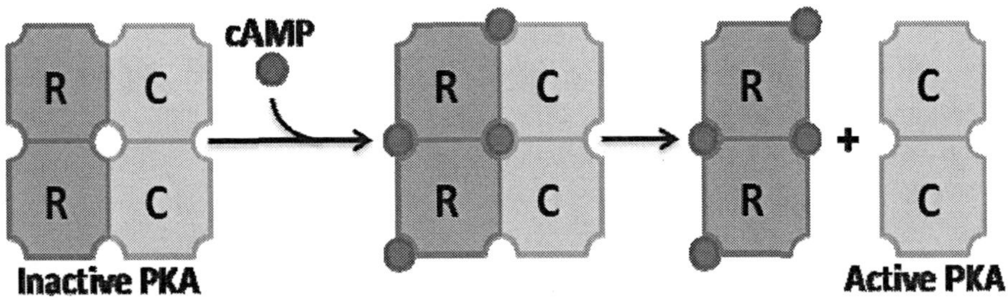

Figure 34. Activation of cAMP-dependent protein kinase (PKA) through dissociation of the two regulatory subunits (R) induced by cAMP to release the two active catalytic subunits (C).

Control of Rate of Synthesis (Enzyme Induction and Repression) or Degradation of Enzyme Molecules

This is a long-term regulation of the enzyme activity to suite the metabolic state and/or the developmental phase of the cell. Control of the rate of synthesis and proteolytic degradation of the enzymes are comparatively slow mechanisms for regulating enzyme concentration; with response times of hours, days or even weeks. To restrict the enzyme activity to such metabolic state and/or developmental phase, the enzyme must be degraded by proteolytic enzymes in a regulated fashion. Regulation of enzyme synthesis is mainly by controlling rate of its gene transcription, its mRNA half-life and/or rate of translation into the protein. Rate of degradation is controlled by complex signaling process and by inherent stabilizing/destabilizing factors in the protein structure itself.

Genes of some enzymes may be expressed at a constant rate whatever the cell metabolic or developmental state and are called *housekeeping or constitutional enzymes.* Their availability is thus controlled mainly by their rate of degradation. They are named so because they perform essential life-

incompatible general housekeeping function in every cell, e.g., enzymes of central metabolic pathways. Other enzyme encoding genes may be induced or repressed/derepressed to produce more enzyme molecules to meet cellular conditional requirements. The induction could be hormone- (e.g., metabolic hormones; insulin, glucagon, thyroxine, catecholamines and glucocorticoids) or substrate-dependent.

Inducible or derepressible enzymes are commonly observed in metabolism of all forms of life. Availability of specific substrate increases rate of synthesis of certain enzyme(s) required for substrate assimilation. Example is the derepression of the β-galactosidase gene (Lac-Operon in general) in *E. coli* bacterium in absence of glucose and presence of lactose. In this respect, unmetabolizable gratuitous inducer (e.g., isopropylthiogalacoside as compared to lactose for β-galactosidase) or derepressor compound is similar in structure to the inducer and is able to induce the enzyme synthesis. One inducer may induce synthesize of a number of enzymes at the same time that is called "coordinate induction". Many toxins/drugs that enter the body induce the rate of synthesis of enzymes that detoxify them to a rate that may reach 100-times higher than the normal basal level.

The genes of other enzymes may be repressed to reduce an existing high rate of synthesis of an enzyme. This could be due to accumulation of the product of an enzyme, or availability of an alternative preferred substrate. Certain bacteria are able to synthesize a particular amino acid and the accumulation of such amino acid decreases rate of synthesis of enzyme(s) synthesizing that amino acid. In this case the amino acid is called co-repressor because it binds and activates a repressor transcription factor to reduce gene(s) expression. Removal of the co-repressor amino acid de-represses (i.e., induces) these gene(s) again. This phenomenon is not restricted to bacteria, but operates to all synthetic pathways in the body, too. Cholesterol and its derivatives are strong repressors for the expression of the key regulatory enzymes for cholesterol synthesis. Thus, induction/derepression and repression are carried out by direct or indirect modulation of the activity of a group of specific DNA sequence binding proteins called transcription factors.

Zymogens (or Proenzymes)

Most enzymes particularly those functioning extracellularly and in body lumens, e.g., digestive enzymes are synthesized in an inactive form called zymogens or proenzymes, e.g., the pancreatic proproteases (trypsinogen, chymotrypsinogen, proelastase and procarboxypeptidase), gastric pepsinogen and blood clotting and clot dissolution factors (enzymes).

Zymogens are inactive due to presence of an inhibitory extra-polypeptide chain, a conformational restraint, requirement of an activating protein, and/or presence of inhibitory subunit. These factors block the active sites of the enzyme so as not to manipulate the substrate. The release of conformational restraint is exemplified by the activating action of gastric HCl on pepsinogen. Proteolytic cleavage of peptide chain is exemplified by activation of blood clotting factors; trypsinogen activation into trypsin by enteropeptidase in the duodenum; trypsin activation of proelastase, chymotrypsinogen and procarboxypeptidase; and, autoactivation of pepsinogen by pepsin and trypsinogen by trypsin.

Other activations may require; dissociation of regulatory subunit (e.g., cAMP-dependent protein kinase upon binding to cAMP), association with another activating protein (e.g., co-lipase for pancreatic lipase and Ca^{2+}-calmodulin for dependent enzymes, e.g., protein kinases), a cofactor (e.g., Ca2+ for protein kinase C and Mg^{2+} for kinases), coenzyme, allosteric activator (e.g., glucose-6-phosphate for glycogen synthase) and/or a regulatory covalent modification (e.g., phosphorylation/dephosphorylation and acetylation/ deacetylation). Most of the metabolic enzymes are inactive at one metabolic state of the cell and require the aforementioned approaches to attain activation.

The metabolic significance of zymogens include; providing a new mechanism for regulating the enzyme activity by determining where and when to be activated; inactive zymogens avoid their secreting cells and the transporting duct system from their effect; it allows storing them in large amount till needed, e.g., blood clotting; and, allow amplifying the physiological effect.

ISOENZYMES (ISOZYMES)

Isoenzymes are structural isomers of the same enzyme isolated from different tissues or subcellular compartments of the same tissue in a single organism. Enzymes with diverse structural differences that catalyze the same reaction using same substrate(s) and producing same product(s) isolated from different species are called *allozymes*. Isoenzymes are physically (amino acid sequence, structurally, electrophoretically and immunologically) distinct forms of the same enzyme that catalyze the same chemical reaction(s) one same substrate(s) to produce same product(s). Isoenzymes differ in their catalytic activity (reaction direction, K_{cat}, K_m and V_{max}), in distribution between different tissues and subcellular compartments, coenzyme/cofactor/prosthetic group requirement, regulation (including sensitivity to inhibitors, and other inactivators, e.g., heat liability), usage of alternative substrates and metabolic role. *The physical differences between isoenzymes may come from;* i) different expressing gene alleles (with same chromosomal locus) or genes (with different chromosomal loci), ii) different composing subunits, and/or, iii) different posttranslational modification of composing subunits. It is possible to correlate the intracellular location of isoenzymes with their metabolic functions, e.g., cytosolar vs. mitochondrial isozymes of each of isocitrate dehydrogenase, malate dehydrogenase and phosphoenol pyruvate carboxykinase.

About half of the known enzymes exist as isoenzymes. Isoenzymes have greatly expanded our understanding of the metabolic regulation. They provided us with a multitude of highly sensitive biomarkers of normal and altered differentiation and development, e.g., during carcinogenesis and teratogenesis – other than their role in diagnostic clinical biochemistry.

EXTRA ATTENTION: ENZYME FAMILIES

An enzyme family is a group of enzymes with a unique active site three-dimensional structure and essential residues and identical mechanism of reaction but differ in their substrate specificity, tissue distribution, regulatory, functional, pharmacological and some structural properties. An example of an enzyme family is the serine proteases that include; typsin, chymotrypsin, elastase, thrombin (EC 3.4.21.5) and subtilisin. These enzymes possess superimposable active site 3D-structure and same essential residues known as the charge relay triad (Ser, His and Asp). Consequently, they possess the same mechanism of proteolytic reaction. Another example is the dehydrogenases with superimposable redox domain and (NAD/P^+) binding domain characterized by $\beta\alpha\beta$ domain, i.e., α-helix flanked by two β-sheets. Examples of within this family include alcohol, lactate, malate, and glyceraldehyde-3-phsphate dehydrogenases.

BIOCHEMICAL SIGNIFICANCE OF ISOENZYMES INCLUDES EXPLAINING

- The enzyme structure-function relationship and enzymatic mechanisms, e.g., usage of alternative substrates and differential effects of inhibitors on different isoenzymes.
- The metabolic differences among the subcellular organelles, e.g., the cytosolar (NADPH-dependent) vs. the mitochondrial (NADH-dependent) isocitrate dehydrogenases.
- The tissue-specific differential expression of genes and dependent tissues metabolic differences, e.g., different fate of lactate in different tissues.
- The individual differences in nutrients, drugs and toxins metabolism, e.g., rapid vs. slow acetylators.
- The genetic bases of some inborn errors of metabolism.
- The utility of enzymes in laboratory clinical diagnostic application.

For example, serum contains five electrophoretically distinct *Lactate Dehydrogenase* (LDH) isoenzymes; namely LDH-1, LDH-2, LDH-3, LDH-4 and LDH-5 (See Table 4). A sixth, atypical LDH isoenzyme was found in male genital tissues, called LDHx. They catalyze the same reaction, i.e., the reversible interconversion of lactate and pyruvate.

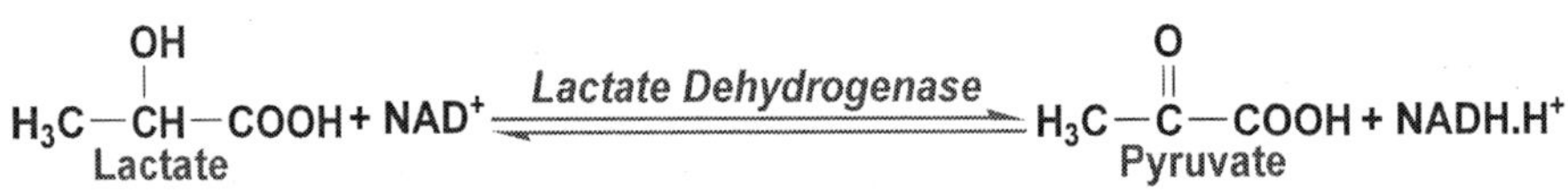

LDH is a tetramer consisting of two types of polypeptide chains designated as M and H. M is referring to the predominant skeletal muscle LDH-5 (MMMM) and H refers to the predominant heart LDH-1 (HHHH). Therefore, these isoenzymes differ physically, in their quaternary structure, the catalytic activity (K_m), metabolic role, pH optima, heat liability, diagnostic value, sensitivity to inhibitors and usage of alternative substrates. The difference in electrophoretic mobility is due to different electric charges of the isoenzymes due to difference in composing subunits.

LDH-1 has the highest negative charge and fastest electrophoretic mobility because of its higher proportion of aspartate and glutamate than the other forms, whereas, LDH-5 is the slowest moving fraction. LDH-4 and

LDH-5 are heat labile, whereas, LDH-1 and LDH-2 are relatively heat resistant even at 60 °C. They differ also in their sensitivity to inhibition by urea, where, hepatic LDH-5 is inhibitable. Also, cardiac LDH-1 and -2 utilize oxo-butyrate preferentially to pyruvate as alternative substrate, whereas, liver LDH-5 and -4 are relatively less active on oxo-butyrate.

Catalytically, the skeletal isoenzyme (M_4) with high affinity to pyruvate favors formation of lactate from pyruvate to regenerate the limited amount of cytosolar NAD^+ necessary for continuation of anaerobic glycolysis; whereas, the heart isoenzyme (H_4) with low affinity to pyruvate and within good aerobic environment favors formation of pyruvate from lactate. This has a physiological importance in disposing and detoxifying lactate to prevent its building up in plasma (lactatemia and lactic acidosis).

Table 4. The major types of LDH isozymes, structure and their tissue distribution.

Type	Structure	Electrophoretic mobility	Tissue distribution
LDH-1	(H_4) HHHH	Fastest moving	RBCs and heart
LDH-2	(H_3M) HHHM	Follows LDH-1	RBCs and heart
LDH-3	(H_2M_2) HHMM	Follows LDH-2	Brain and kidney
LDH-4	(HM_3) HMMM	Follows LDH-3	Liver and muscles
LDH-5	(M_4) MMMM	Slowest moving	Liver and muscles

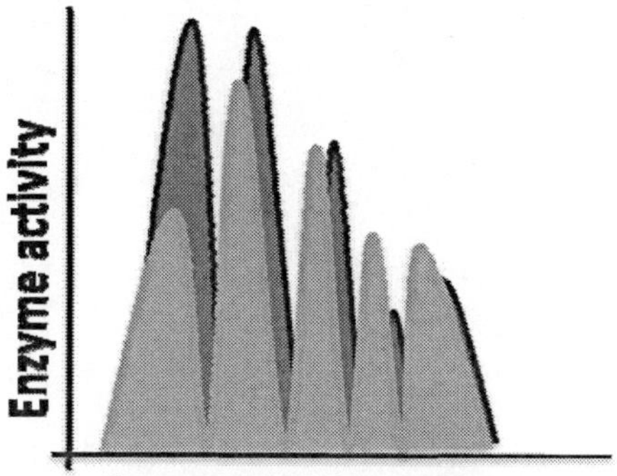

Figure 35. LDH isoenzymes. Electrophoretogram of LDH isoenzymes detected as enzyme activity in a healthy individual (blue shade) and in a patient with acute myocardial infarction (red shade).

Cellular damage of skeletal muscles, myocardium or liver causes increase in total serum LDH particularly the predominant isozyme - easily identifiable by electrophoresis. In normal serum, LDH-2 is the most prominent isozyme, whereas, the slowest peak of LDH-5 is rarely seen. After myocardial infarction, the faster isoenzymes LDH-1 and LDH-2 predominate (Figure 35; presents an electrophoretogram for serum proteins labeled for lactate dehydrogenase in these three conditions). In acute viral hepatitis, the slowest isoenzymes LDH-5 and LDH-4 predominate.

Total serum LDH is frequently elevated in neoplastic diseases with a pattern shifting towards slower migrating components (LDH-3, 4 and 5). LDH-5 increases in breast carcinoma, malignancies of CNS, and prostatic carcinoma. LDH-2 and -3 levels rise in leukemias. LDH-2, -3 and -4 levels increase in cancers of testes and ovary.

Chapter VIII

The Basic Principle of Enzyme Extraction and Kinetic Characterization: Tyrosinase as an Example

Introduction

Proteins - including enzymes - are differentially soluble in salt solutions, and enzyme extraction procedures often begin with salt precipitation; typically, ammonium sulfate. On the simplest level, proteins can be divided into albumins and globulins on the basis of their solubility in dilute salts (salting in). Albumins are considered to be soluble while globulins are insoluble. However, the salt concentration required for solubilization or precipitation is relative within each of these two major groups, and as the salt concentration is increased, most proteins will precipitate (salting out). After homogenizing the tissue source of an enzyme into a solution that retains the target enzyme in its soluble state, the enzyme can be subsequently separated from all insoluble proteins by centrifugation or filtration. However, in such crude preparation the enzyme will be impure - being contaminated with several other soluble proteins. Subsequently, the serial addition into such solution of fine aliquots of a concentrated ammonium sulfate solution precipitates individual proteins according to their differential solubility. Such enzyme preparation is purer, or enriched for a given enzyme. Although salts can cause denaturation of the three-dimensional structure of the enzyme, the effects of ammonium sulfate are usually reversible, e.g., by dialysis. More absolute purity of an enzyme extract requires the subjection of such fraction to further purification procedures, e.g., electrophoresis and/or ion-exchange and affinity column chromatography. Monitoring the specific enzyme activity in preparations at each step of extractions helps determining the effectiveness of the purification – see later for approaches for enzyme assays.

As an Example, the Procedures for Extracting and Kinetically Characterizing the Tyrosinase Enzyme from Potatoes will be Explained

Introduction

Tyrosinase (EC 1.14.18.1) is also called monophenol monooxygenase, phenolase, monophenol oxidase, cresolase - particularly when isolated from plant sources - and functionally is an oxygen oxidoreductase enzyme. Tyrosinase is a copper-containing enzyme that catalyzes the rate-limiting step in melanin biosynthesis, the hydroxylation of L-tyrosine to 3,4-dihydroxy-L-phenylalanine (L-DOPA) and the subsequent oxidation of L-DOPA to L-dopaquinone- thus it works as a hydroxylase and oxidase. In the absence of thiol compounds L-dopaquinone undergoes a rapid oxidation and spontaneous rearrangement leading to L-dopachrome, and ultimately, to the melanin polymer. Dopaquinone is an intermediate metabolite in the production of melanin and other plant pigments responsible for blackening sliced tuber and fruits exposed to air. It is widespread in fungal, plants and animals tissues. Tyrosinases from different species are diverse in their structural properties, tissue distribution and cellular location. The enzymes found in plant, animal and fungi tissue frequently differ with respect to their primary structure, size, glycosylation pattern and activation characteristics. However, all tyrosinases have in common a binuclear type 3 copper center within their active site, where, two copper atoms are each coordinated with three histidine residues. The two copper atoms within the active site of tyrosinase enzymes interact with dioxygen to form a highly reactive chemical intermediate that then oxidizes the substrate. In animal cells majority of the enzyme are particle-bound to microsomes and melanosomes, and, 20% is soluble – prepresenting two forms of the enzyme. Human tyrosinase is a single membrane spanning transmembrane glycoprotein and the catalytically active domain of the protein resides within melanosomes, whereas, a small enzymatically non-essential part of the protein extends into the cytoplasm of the melanocyte.

Several approaches for extraction were employed according to the source tissue, e.g., fungal mycelia of N. crassa were first liquid nitrogen frozen, homogenized with a French Press while frozen, proteins were precipitated in ammonium sulfate, and the enzyme was purified chromatographically on Sephadex and Celite columns. When hamster melanomas were the source, this technique was modified by addition of acetone extractions as well as DEAE-cellulose chromatography and alumina treatments. The simplest technique was used upon extraction from plant tissues based principally on ammonium sulfate precipitation of proteins. During the synthesis of melanin pigment, the enzyme catalyzes the conversion of tyrosine + $\frac{1}{2}O_2$ $\Rightarrow$ dihydroxyphenylalanine (DOPA), and then catalyzes conversion of 2 DOPA + O_2 $\Rightarrow$ into 2 dopaquinone + 2 H_2O. Dopaquinone is spontaneously converted into dopachrome, a dark orange pigment (dopaquinone $\Rightarrow$ Leukodopachrome + dopaquinone $\Rightarrow$ Dopachrome + DOPA). Therefore, the catalytic activity of the enzyme will be monitored through spectrophotometric determination of produced dopachrome with an absorbance maximum at 475 nm from the substrate DOPA (DOPA + $\frac{1}{2}O_2$ $\Rightarrow$ Dopachrome).

STEP 1: ENZYME EXTRACTION

Materials

Source material: Potatoes; *Solutions:* 0.1 M sodium fluoride (NaF – toxic treat with caution), 0.1 M citrate buffer, pH 4.8; and saturated ammonium sulfate (4.1 M at 25 °C); and *Tools:* rubber gloves, volumetric cylinders (50 mL, 100 mL, 250 mL), cheesecloth, beakers (100 mL, 250 mL), chilled centrifuge tubes (30 - 50 mL), refrigerated centrifuge, glass stirring rod, paring knife, and blender.

Procedure

In the blender, add 100 gm potato (2 cm square pieces) from a peeled potato + 100 ml of sodium fluoride and homogenize for ~1 minute at high speed. Sieve the homogenate through several layers of cheesecloth and into a beaker. Centrifuge aliquots at 300xg for 5 minutes at 4 °C to get rid of any course precipitates. Add equal volumes of the saturated ammonium sulfate and the supernatant - tyrosinase is insoluble in 50% ammonium sulfate. Proteins will precipitate as a white flocculent. Centrifuge aliquots of the treated homogenate in the chilled centrifuge tubes at 1,500xg for 5 minutes at 4 °C. Discard the supernatant fluid and add 60 mL citrate buffer onto the collected pellet in 100 mL beaker and stir the contents well for 2 minutes to break up the pellet on ice. Centrifuge aliquots again at 300xg for 5 minutes at 4 °C and recover the supernatant which contain the solubilized tyrosinase extract - it is soluble in the citrate buffer and place on ice. In this condition, the enzyme is stable for ~1 hour.

Step 2: Preparation of Standard Curve and Calculation of the Extinction Coefficient

The standard curve could be prepared from known pure enzyme preparation or from the product - as is the case in this procedure.

Materials

Solutions: the solubilized enzyme extract, 8 mM L-DOPA and 0.1 M citrate buffer, pH 4.8; and, *Tools:* Test tubes, 5 mL pipette, spectrophotometer and cuvettes.

Procedure

Preparation of the stock standard dopachrome (mix 10 mL of the colorless 8 mM DOPA + 0.5 mL of the enzyme extract and leave to stand for 15 minutes at room temperature ⇒ 8 mM dopachrome because all of the L-DOPA are converted into dark orange dopachrome within this time). Safe dopachrome in dark brown bottle or away from light because it is light sensitive. Serially dilute the stock standard (8.0 mM) 1:1 into 6 tubes (#2 - 7) containing 1.5 mL buffer by adding 1.5 mL of the higher concentration. Tube #1 will contain 3 mL of buffer to be used as the black tube, and tube #8 will contain 3 mL of the stock solution. Therefore, there will be 8 standards 0.0, 0.125, 0.25, 0.5, 1.0, 2.0, 4.0 and 8.0 mM dopachrome concentration - corresponding to 0.0, 0.375, 0.75, 1.5, 3.0, 6.0, 12.0, 24 micromoles of dopachrome amount per the 3 mL-tube. Set the spectrophotometer to a wavelength of 475 nm and zero it by the blank solution. Read the absorbance of each of the solutions in tubes 2-8 and register data as tabulated below.

Tube #	**Dopachrome Concentration (mM)**	**Absorbance, 475 nm**	**Extinction CoefficientA/C)**
1	0		
2	0.125		
3	0.25		
4	0.5		
5	1.0		
6	2.0		
7	4.0		
8	8.0		
Average Extinction Coefficient			

The average extinction coefficient is used in subsequent determinations of dopachrome concentrations as a function of the tyrosinase activity in similar conditions without requiring prior preparation of such standard curve again. However, more accurate extinction coefficient is extracted from the linear regression analysis of the standard curve, and computing the slope and y intercept. The slope of the linear regression represents the extinction coefficient. The molar extinction coefficient of dopachrome is 3600.

The straight line standard curve is developed by plotting the absorbance values on the y axis against corresponding concentrations of dopachrome on the x axis and straightly connecting the developed intercepts. The equation for a straight line is $y = mx + b$, where m is the slope of the curve and b is its y intercept. Since substrate and product are in a 1:1 ratio in this reaction, the amount of product formed equals the amount of substrate used and both proportionate directly with the optical density of dopachrome as intensity of the orange color formation in solution measured at 475 nm.

Step 3: The Effect of Enzyme Concentration on the Rate of the Reaction

Materials

Solutions: Enzyme extract, 0.1 M citrate buffer, pH 4.8, and 8 mM L-DOPA, and, *Tools:* 10 mL pipette, spectrophotometer and cuvettes, stopwatch and ice bath.

Procedure

To determine the kinetic effects of the enzyme reaction, first determine an appropriate dilution of your enzyme extract that gives a rate of reaction of 5 - 10 μM L-DOPA conversion/minute. From the enzyme extract prepared as above, prepare a serial dilution by placing 9.0 mL of citrate buffer into each of three test-tubes. Label the tubes 1/10, 1/100 and 1/1000 besides the original extract. Dispense 1.0 mL of your enzyme extract into tube 1/10 and mix by inversion, then take 1.0 mL of it and add into tube 1/100, and mix by inversion, then take 1.0 mL of it and add into tube 1/1000 mix by inversion. All solutions should be kept on ice. Zero the spectrophotometer absorbance using 2.5 mL of citrate buffer + 0.5 mL of enzyme extract. Dispense 2.5 mL of 8 mM L-DOPA to each of 4 tubes, so as each contain 20 μM L-DOPA (8 mM L-DOPA/L = 8 μM L-DOPA/mL X 2.5 mL = 20 μM L-DOPA/tube). Add 0.5 mL of original enzyme extract to one of the 4 tubes and mix by inversion. Put into the spectrophotometer and immediately begin timing the increase in absorbance due to the progressive conversion of L-DOPA into dopachrome to measure the time required for the conversion of 8 μM/tube L-DOPA into 8 μM/tube dopachrome, i.e., 2.67 μM/mL (or 2.67 mM/L). apply this concentration into the standard curve developed above to check for the absorbance corresponding to 2.67 mM dopachrome. This absorbance value will be the end point for the reaction at which the time elapsed should be 3 -5 minutes. Earlier time points means that enzyme concentration is too high and time is too fast for developing the linear portion of the enzyme-substrate curve. In this case the experiment should be repeated using the diluted enzyme preparations (1/10, then 1/100, then 1/1000). At that point, to calculate the reaction velocity (rate of activity) divide the amount of the product (2.67 mM/L)/minutes (3-5) and multiply by the dilution factor (if the diluted preparation is used; i.e., 10, 100 or 1000) to get rate of activity in μM/minute/0.5 mL enzyme preparation. Substrate concentrations are roughly calculated because of the diluting effect of the enzyme solutions added.

STEP 4: THE EFFECT OF PH ON THE RATE OF THE REACTION

Materials

Solutions: 8 mM L-DOPA in citrate buffer adjusted to pH values of 3.6, 4.2, 4.8, 5.4, 6.0, 6.6, 7.2, 7.8, and enzyme extract, and *Tools:* spectrophotometer and cuvettes and a stopwatch.

Procedure

Dispense 2.5 mL of the 8 mM L-DOPA solutions at the different pHs in each of 8 tubes labeled 3.6, 4.2, 4.8, 5.4, 6.0, 6.6, 7.2, and 7.8. Taking one tube at a time, add 0.5 mL of the diluted enzyme extract used in step 3 (that converted 2.67 μM/minute L-DOPA), and mix by inversion. Start timing the reaction and insert into the spectrophotometer and stop timing at the time required to reach the same absorbance as in step 3. Repeat the same procedure for each of the remaining DOPA tubes and register the developed data into the following table and plot pH values on x axis vs. the calculated reaction Velocity on the y axis.

pH/tube	Time (Minutes)	Dopachrome	Velocity (Micromoles/Minute)
3.6		2.67 μM	
4.2		2.67 μM	
4.8		2.67 μM	
5.4		2.67 μM	
6.0		2.67 μM	
6.6		2.67 μM	
7.2		2.67 μM	
7.8		2.67 μM	

Step 5: The Effect of Temperature on the Rate of the Reaction

Materials

Solutions: enzyme extract and 8 mM L-DOPA in citrate buffer adjusted to pH value of 6.6, and *Tools:* incubators or water baths adjusted to 10, 15, 20, 25, 30, 35 and 40° C (or temperature-controlled spectrophotometric chamber), spectrophotometer and cuvettes, and stopwatch.

Procedure

Dispense 2.5 mL of the 8 mM L-DOPA solution into each of 7 tubes labeled 10, 15, 20, 25, 30, 35, and 40 °C and incubate each at the corresponding temperature. Dispense 0.5 mL of the diluted enzyme extract (that yields 2.7 μM dopachrome/minute) to each of a second set of 7 tubes labeled 10, 15, 20, 25, 30, 35, and 40 °C and incubate each at the corresponding temperature. Allow all of the tubes to temperature equilibrate for 5 minutes. Zero the spectrophotometer at 475 nm with the buffer containing the enzyme in the same proportions (2.5 + 0.5 mL) and mix the two 10 °C tubes and place in the spectrophotometer and begin timing the reaction till reaching the absorbance end point equivalent to the conversion of 2.7 μM L-DOPA. Repeat with the other 6 double sets of tubes and register the developed data in the following table and plot temperature values on x axis vs. the calculated reaction Velocity on the y axis. This experiment could be repeated with temperature increasing by 1.0 °C for a range of 20 - 40 °C.

Temperature °C	Time (Minutes)	Dopachrome	Velocity (Micromoles/Minute)
10		2.67 μM	
15		2.67 μM	
20		2.67 μM	
25		2.67 μM	
30		2.67 μM	
35		2.67 μM	
40		2.67 μM	

STEP 6: DETERMINATION OF K_M AND V_{MAX} OF THE REACTION

Materials

Solutions: enzyme extract and 8 mM L-DOPA in citrate buffer adjusted to pH 6.6, and *Tools:* spectrophotometer and cuvettes and stopwatch.

Procedure

Serially dilute the 8 mM L-DOPA standard into; 0.5 mM, 1 mM, 2 mM 4 mM, and 8 mM in the buffer. In a series of 5 tubes dispense 2.5 mL of each concentration. Add 0.5 mL of the properly diluted enzyme solution to each tube one at a time and read absorbance at room temperature for all tubes when the end point absorbance is reached and calculate reaction velocity (v; μM/minutes). Calculate the 1/s and 1/v of each reaction and register data as follows. Plot the rate of L-DOPA conversion (v) values on the x axis against substrate concentration on the y axis to get the Michaelis-Menten plot. A linear increase of the absoebance is expected in the first 2 minutes followed by progressive decreases in the oxidation rate. Plot the double reciprocal of the values, i.e., 1/s vs. 1/v to get the linearized Lineweaver-Burke plot. Perform a linear regression analysis on the second plot and compute the slope and both y and x intercepts. The x intercept is $-1/K_m$, the negative inverse of which is the Michaelis-Menten Constant. The y intercept is $1/V_{max}$ and the slope equals Km/V_{max}. The K_m is expected to be 250 μM on L-DOPA as a substrate.

L-DOPA Concentration (mM)	Velocity Micromoles/Minute	1/s	1/v
0.5		2.00	
1.0		1.00	
2.0		0.50	
4.0		0.25	
8.0		0.125	

Step 7: Determination of the Effect and Types of the Enzyme Inhibitors

Materials:

Solutions: enzyme extract, 8 mM L-DOPA, 8 mM benzoic Acid, 8 mM KCN, and 0.1 M citrate buffer, pH 6.6.

Procedure:

Copper chelators, benzoic acid (competitive inhibitor for the first substrate; L-DOPA) and cyanide (competitive inhibitor for the second substrate; O_2) inhibit tyrosinase. To determine the inhibitory effects of benzoic acid and cyanide, set up a series of 11 tubes (# 1 – 11) for each inhibitor and add 8 mM L-DOPA, inhibitor and buffer volumes in mL as indicated in the following table. L-DOPA final concentrations within the reaction will be decreasing from, 6.67, 6.0, 5.33, 4.67, 4.0, 3.33, 2.67, 2.0, 1.33, 1.67, 0.0 mM.

Tube #	8 mM L-DOPA, mL	8 mM Benzoic Acid, or 8 mM Potassium cyanide, mL	Buffer, mL
1	2.0	0.5	0
2	1.8	0.5	0.2
3	1.6	0.5	0.4
4	1.4	0.5	0.6
5	1.2	0.5	0.8
6	1.0	0.5	1.0
7	0.8	0.5	1.2
8	0.6	0.5	1.4
9	0.4	0.5	1.6
10	0.2	0.5	1.8
11	0	0.5	2.0

Using one tube at a time, add 0.5 mL of the properly diluted enzyme solution (that yields 2.7 μM dopachrome/minute) and determine the time required to convert reach the expected OD endpoint and calculated the reaction velocity μM/minute. Calculate 1/s and 1/v values form s and v for each tube. Plot 1/v vs. 1/s for each inhibitor and calculate the V_{max} and K_m for the presence of each inhibitor. Determine whether these inhibitors are competitive (benzoic acid vs. L-DOPA), non-competitive (potassium cyanide vs. L-DOPA whwere it is reversed by addition of copper II 50 μM after dialysis) or uncompetitive. The experiement could be repeated the same with L-phenylalanine or L-tyrosine as examples of competitive, and, 2,9-dimethyl-1,10-phenanthroline as uncompetitive inhibitors of L-DOPA oxidation.

Step 8: Protein Concentration/ Enzyme Activity

Materials

Solutions: commercially pure tyrosinase 0.7 μg/4 mL in 0.1 M citrate buffer, pH 6.6, L-DOPA 4 mg/mL in 0.1 M citrate buffer, pH 6.6, and Lowry or Bradford Protein determination reagents, and *Tools:* UV spectrophotometer.

Procedure

Measure the OD at 280 nm the enzyme extract (or measure total protein content by Lowry or Bradford Protein determination reagents) and dilute with the buffer to 0.7 μg/4 mL. Temperature equilibrates the two enzyme solutions at 30 °C for 5 minutes. Zero the spectrophotometer at 475 nm with citrate buffer as the blank. Add 1.0 mL L-DOPA solution to 4 mL of the commercial enzyme preparation and read absorbance immediately at 475 nm then incubate again for 5 minutes and read absorbance again. Multiply net the absorbance (A2- A1) by 3.7×10^4 (the molar absorbance coefficient for dopachrome) and divide by 5 to calculate the specific activity of the commercial enzyme preparation; μM dopachrome/ minute/mg protein. Repeat the same steps for the extracted enzyme and calculate its specific activity. The specific activity is converted into units of activity activity/mg protein of the enzyme preparation, where 1 unit of enzyme activity (1 unit transforms 1 μM of substrate /minute under define conditions) causes 0.81 changes in absorbance under conditions specified in this experiment.

Chapter IX

MEASUREMENT OF ENZYME ACTIVITY (ENZYME ASSAY)

INTRODUCTION

Enzyme assays are laboratory methods for measuring enzymatic activity that are vital for the study of enzyme kinetics and enzyme inhibition; along with research and laboratory diagnostic applications. The measurement of individual enzyme activity does not require prior purification because it can be conducted on complex samples, e.g., body fluids and tissue homogenates. This utilizes the specific enzyme action on its specific substrate under optimized reaction conditions.

Because the reaction progression curve is not linear, the maximum slope is located at the nearly linear part (15-20% of the total reaction change) very close to the very fast reaction start point (Figure 36). At this area the reaction rate is called *the initial rate of the reaction* (V_o) where the curve is steepest. The initial rate of the reaction equals the slope of the tangent to the curve as closest to the time 0 as feasible, or, equals the amount of the product formed in the initial few seconds. *The later decrease in the rate of the reaction progression and hence the slope is reasoned to;* i) lowered substrate availability, ii) increased rate of the reverse reaction towards its equilibrium, iii) lowered catalysis due a product-dependent change in pH, iv) feedback inhibition by the accumulating product, and, vi) time-dependent inactivation of the enzyme, and v) occupancies of the available active sites of the enzyme.

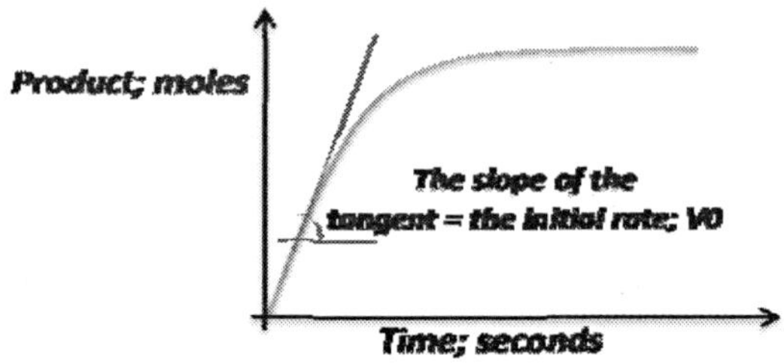

Figure 36. Measurement of enzyme activity is particularly noticed at the initial nearly linear progress of the reaction where the slope of the tangent to the curve equals the initial rate.

Unit of Serum Enzyme Activity

It is difficult to measure the amount of enzyme in the conventional units of mass or moles like any other chemical. Reaction rate is an accepted expression of enzyme activity. Specific activity is number of enzyme units/mg enzyme protein. An enzyme unit is the amount of the enzyme that catalyzes the transformation of 1 µM amount of substrate/minute at 30 °C under optimal chemical environment (optimal pH and unlimiting substrate concentration). Because most of the enzyme preparations are not absolutely pure, enzyme activity and protein content do not mach due to other contaminating proteins without the specific enzymatic activity. Therefore, to determine Specific activity both protein content and enzyme activity are required to be measured by two different procedures; spectrophotometrically and kinetically, respectively. Enzyme activity = moles of substrate converted per unit time = rate X reaction volume. *One international unit (IU) of enzyme activity* is the activity of the enzyme which transforms one µmole of substrate per minute under specific conditions and at defined temperature (mostly 30 ^{o}C), and is expressed as IU/mL. The *SI Katal unit of enzyme activity* is the amount of the enzyme that converts 1 mole of substrate into product in 1 second that is an excessively large unit, whereas, a nanokatal equals 0.06 IU. *The specific activity of an enzyme preparation* is the catalytic enzyme activity (in nanokatal or IU) in 1 mg of total protein in the enzyme preparation. It is expressed in µmol/min^{-1} mg^{-1} and reflects the purity of the enzyme preparation.

Enzyme assays measure either the consumption of substrate or the production of product over time by using one of four methods; the initial rate, progress curve, transient kinetics and the relaxation assays:

- *Initial rate assay* in which the rate is measured during a very short period after mixing the enzyme with a large excess of the substrate, where the enzyme-substrate intermediate builds up in the fast initial transient. After the attainment of the linear-steady state, typically the accumulation of product with time is monitoring. The initial rate assay is the simplest to perform and analyze, because it is relatively free from complications such as back-reaction and enzyme degradation. Therefore, by far it is the most commonly used type of experiment in enzyme kinetics.
- *Progress curve assay* where the concentration of the substrate or product is recorded as a function of time after the initial fast transient and for a sufficiently long period to allow the reaction to approach equilibrium. Progress curve assay was widely used in the early period of enzyme kinetics.
- *Transient kinetics assay* where the reaction behavior is tracked during the initial fast transient as the intermediate reaches the steady-state kinetics period. This assay is difficult to perform than either of the above two classes because it requires rapid mixing of reagents and observation techniques.
- *Relaxation assay* that considers a fully reversible reaction at the steady-state equilibrium of enzyme, substrate and product, then, the equilibrium is perturbed through, e.g., sharp change in temperature, pressure or pH, then, the return to equilibrium is monitored. This assay is not typically used for kinetic studies because it is relatively insensitive to mechanistic details.

According to the reaction follow up method, enzyme assays are of two types; continuous assays, where the assay gives a continuous reading as a function of the enzyme activity, and, discontinuous (endpoint) assays that requires intermittent or final stoppage by taking samples of the reaction to monitor its progress.

Continuous assays directly reflect the progress of the reaction because of the nature of the monitoring method. Types of the monitoring methods include; spectrophotometric, fluorometric, microcalorimetric, chemiluminescent (chemo- and bioluminescence), light scattering, electrochemically, and gasometrically.

- *Spectrophotometric assay* follows the course of the reaction through measuring the change in the light absorbance (ultraviolet or visible colorimetric) of the assay solution. The 3-(4,5-Dimethylthiazol-2-yl)-2,5-diphenyltetrazolium bromide (MTT) cell toxicity assay that monitors succinate dehydrogenase oxidation of tetrazolium dye as a mitochondrial activity indicator is an example of a direct colorimetric assay. Ultraviolet (UV) light absorbance at 340 nm is used with oxidoreductase coupled reactions that generates or consumes NADH and NADPH as coenzymes. The change of absorbance could come out of the target reaction or from a coupled reaction that uses the target reaction's product as a substrate. The later couple assay method requires the presence of reactants and required factors for the target (or initial) enzyme and the additional enzymes (auxiliary or intermediate, and, indicator or final). For example, during the assay of aspartate transaminase (AST; EC 2.6.1.1), the enzyme malate dehydrogenase (EC 1.1.1.37) is used as the indicator enzyme where it converts oxaloacetate produced from the initial AST reaction into malate on the expense of consuming NADH.H$^+$ into NAD$^+$ that correlates reduction in 340 nm UV absorption.

$$\textbf{Aspartate} + \alpha\textbf{-ketoglutarate} \xrightleftharpoons[PLP]{\textit{Aspartate Transaminase}} \textbf{Oxaloacetate} + \textbf{Glutamate}$$

$$\textbf{Oxaloacetate} + \textbf{NADH.H}^+ \xrightleftharpoons{\textit{Malate Dehydrogenase}} \textbf{Malate} + \textbf{NAD}^+$$

Another example, creatine kinase (CK; 2.7.3.2) is measured using hexokinase (EC 2.7.1.1) as auxiliary enzyme to convert ATP produced by the initial CK reaction into ADP with activation of glucose into glucose-6-phosphate. The later is converted by the indicator glucose-6-phosphate dehydrogenase (EC 1.1.1.49) into 6-phosphogluconolactone on the expense of increasing NADPH.H$^+$ from NADP$^+$ that correlates increases in 340 nm UV absorption.

$$\textbf{Creatine phosphate} + \textbf{ADP} \xrightleftharpoons{\textit{Creatine Kinase}} \textbf{Creatine} + \textbf{ATP}$$

$$\textbf{Glucose} + \textbf{ATP} \xrightarrow{\textit{Hexokinase}} \textbf{Glucose-6-phosphate} + \textbf{ADP}$$

$$\textbf{Glucose-6-phosphate} + \textbf{NADP}^+ \xrightarrow[\textit{dehydrogenase}]{\textit{Glucose-6-phosphate}} \textbf{Glucose-6-phosphate} + \textbf{NADP}^+$$

b. *Fluorometric assay* monitors the fluorescent light emitted from the assay solution (due to substrate or product) after absorbing light of a different wavelength. It is much more sensitive than spectrophotometric assays, but can suffer from interference caused by impurities and the instability of many fluorescent compounds when exposed to light. Monitoring the generation or consumption of NADH and NADPH coenzymes is an example where their reduced forms are fluorescent and the oxidized forms are non-fluorescent. Synthetic substrates that release a fluorescent dye in an enzyme-catalyzed reaction are also available, such as 4-methylumbelliferyl-β-D-galactoside for assaying β-galactosidase. It is called phosphorescence when the emitted light comes slower at a longer wave length.

c. *Microcalorimetric assay* monitors the released or absorbed of heat from the many chemical reaction involving some change in heat. These assays can be used to measure reactions that are impossible to assay in any other way.

d. *Chemiluminescent assay* monitors the emission of light during the chemical reaction. It is extremely sensitive and uses the enzyme substrate or synthetic substrate or labeling, e.g., the luciferase enzyme activity (isolated from the fireflies), naturally produces light from its substrate luciferin. The weakest light emitted could be captured by photographic film over days or weeks, e.g., the detection of horseradish peroxidase activity as a common method of detecting horseradish peroxidase-labeled antibodies in western blotting protein detection technique. The light source is not a lamp but the reaction itself or a coupled chemical or electrochemical reaction.

e. *Light Scattering assay* measures the concentration depending on the ability of particulate macromolecules in solution to scatter light into a specific angle which vary as they aggregate or dissociate. Hence the measurement quantifies the stoichiometry of the complexes as well as kinetics. Light scattering assays of protein kinetics (e.g., serum proteins, urinary proteins, or antigen-antibody binding) is a very general technique that does not require detection enzymes or chemicals. It has subtypes; i*) turbidimetry* that measures decrease in the intensity of the incident light (due to scattering, reflectance and absorbance) by the particles in the same direction of the incident light, i.e., it is the same as spectrophotometry, and, ii) *Nephelometry* that rather measures the amount of scattered or reflected light at an angle from the direction of the incident light; mostly 90° on it to give highest sensitivity.

f. *Gasometric assay* monitors reactions that produce or consume a gas, e.g., O_2 uptake by L- and D-amino acid oxidases and CO_2 generation by histidine decarboxylase. The change in the pressure and/or volume of the gas is monitored. It is an extremely tedious and difficult method rarely used nowadays; being replaced by electrochemical gas monitoring methods.

g. *Electrochemical assay* utilizes specific electrode to monitor the reaction. The earliest was the glass electrode for pH monitoring in reaction that produce acid or base, e.g., different hydrolytic reactions. Other ion-selective electrodes include cation electrode for detecting ammonia released from, e.g., L- and D-amino acid oxidases; O_2 electrode to detect O_2 released from, e.g., glucose oxidase reaction; CO_2 electrode to detect CO_2 released from decarboxylation reactions, and, redox platinum electrode for detection of redox change in reactions of oxidoreductases.

Discontinuous assays use discontinuous sampling from the reaction mixture at intervals to monitor the product generation or the substrate consumption as a function of the enzyme activity.

a. *Radiometric assay* measure the incorporation of radioactivity into substrates or its release from substrates. The radioactive isotopes most frequently used in these assays are ^{14}C, ^{32}P, ^{35}S and ^{125}I. Since radioactive isotopes can allow the specific labeling of a single atom of a substrate, these assays are both extremely sensitive and specific. It is particularly valuable for polymerization reactions, e.g., synthesis of DNA, RNA and glycogen. However, it was used for most of the basic biochemical discoveries particularly when crude cellular extracts were used. Radioactivity is monitored γ-solid-phase or β-liquid scintillation counters.

b. *Chromatographic assays* measure product formation after separating the reaction mixture into its components by simple paper/thin layer chromatography or high-performance liquid chromatography.

Chapter X

CLINICAL ENZYMOLOGY

INTRODUCTION

Since the tight control of enzyme activity is essential for homeostasis, any malfunction of a single critical enzyme (mutation, overproduction, underproduction or deletion) can lead to a genetic disease - commonly called inborn errors of metabolism. Thus, a lethal illness can be caused by the malfunction of just one type of enzyme out of the thousands of types present in our bodies.

One example is the most common type of phenylketonuria caused by a mutation of a single amino acid in the enzyme phenylalanine hydroxylase, which catalyzes the first step in the degradation of phenylalanine. The deficiency results in build-up of phenylalanine and related unphysiological by-products. This can lead to mental retardation if the disease is untreated early. Another example is when germline mutations in genes coding for DNA repair XPA – XPG enzymes cause hereditary cancer syndromes such as Xeroderma Pigmentosum with increased liability to multiple cancers. Early diagnosis of such defects by, e.g., detection of the enzyme activity and implicated gene mutations is very essential for early intervention.

Diagnostic Clinical Biochemistry

One major item of the diagnostic clinical biochemistry is the investigation of changes in the level of enzymes and their correlation to the differential diagnosis of diseases and to establish cut-off levels for; normal, benign and malignant diseases. These enzymes changes could be followed up in plasma, serum, urine, urine, blood cells or tissue biopsies. Cross-sectional single or longitudinal serial assays of the serum activity of a selected enzyme(s) may support the diagnosis of a specific disease location and/or extent, disease prognosis, recurrence and or monitoring the response to treatment. Thus, detection of the plasma level of an enzyme immunologically (for its protein amount) or colorimetrically (for its activity, preferred) have the following applications:

- *Diagnosis:* As example, high serum creatine phosphokinase (CPK) on the day of a suspected case of myocardial infarction strengthen the diagnosis if ECG changes are doubtful.
- *Differential diagnosis:* e.g., chest pain associates myocardial infarction and pulmonary embolism. Elevated serum glutamate-oxaloacetate transaminase (GOT) and lactate dehydrogenase (LDH) characterizes myocardial infarction, whereas, elevated serum LDH only characterizes pulmonary embolism.
- *Therapeutic follow up and/or early detection of a disease:* Chronic administration of several therapeutics - e.g., antidepressant and anticancer chemotherapies - elevates serum isocitrate dehydrogenase or ornithine carbamoyl-transferase level when they induce minimal hepatotoxicity. Serum glutamate-pyruvate transaminase (GPT) level elevates in sub-clinical early viral hepatitis.

Plasma enzymes are of two sources; plasma-derived or Cell-derived.

- *Plasma-derived enzymes:* They are normally occurring *functional plasma enzymes*. Their field of activity is plasma components and their activity is higher in plasma than in cells, e.g., coagulation and lipoprotein-metabolizing enzymes. Their clinical importance is limited to diseases related to their own synthesis and function; i.e., abnormalities of metabolism of plasma lipoproteins and blood clotting, and the organ function of their synthesizing tissues, e.g., thromboplastin as a liver function test.

- *Cell-Derived enzymes:* Normally they locate to intracellular compartments; i.e., they are *non-functional plasma enzymes*. A very low plasma level normally exists due to normal wear and tear and diffusion through undamaged cell membranes. Gross damage to the cells or abnormal membrane permeability, overproduction of the enzymes or abnormal high cellular proliferation and/or wear and tear may allow their leakage in abnormally high amount into plasma and other body fluids. The amount and nature of the plasma enzyme(s) reflects the extent and nature of the damaged tissue. They are further subdivided into; secretory and metabolic non-functional plasma enzymes:
    1. *Secretory:* They are synthesized and secreted by specialized glands into body lumens mainly for digestion. Their retrograde escape into blood reflects damage in the tissue of their origin, e.g., pancreatic amylase and lipase in pancreatitis.
    2. *Metabolic:* They are intracellular metabolic enzymes and their appearance in the plasma is mainly due to cellular damage among other factors (See later).

Non-functional Plasma Enzymes:

They may be abnormally increased or decreased than the normal level.

Increased non-functional plasma enzymes could be due to increased release and/or impaired clearance.

- *Abnormally increased release* from cells may be due to:
    1. Pathological apoptosis and/or necrosis of cells, e.g., elevated levels of aldolase (EC 4.1.2.13), CK, LDH and GOT in progressive muscular dystrophy.
    2. Increased membrane permeability without gross cellular damage, e.g., elevated levels of GPT in early stage of viral hepatitis.
    3. Increased intracellular enzyme concentration due to:

i. Protein anabolic drugs, e.g., increased synthesis of liver transaminases.
ii. Higher cellular proliferation and increased cell mass as in malignancies, e.g., elevation of alkaline phosphatase (ALP) in osteoblastic bone lesions and hepatobiliary disease, and elevation of acid phosphatase (EC 3.1.3.2) in cancer prostate.

- *Impaired clearance:* As the case for other plasma proteins, enzymes have specific plasma half-life after which they are disposed by cellular reuptake, degradation and/or excretion in bile or urine. Examples include; elevation of serum Leucine AminoPeptidase (LAP) and ALP in obstructive jaundice and, elevation of several enzymes in nephrotic syndrome and renal failure.

Decreased activity of non-functional plasma enzymes could be due to decreased enzyme synthesis, increased enzyme inhibition and/or deficiency of its activating factors.

- *Decreased synthesis of an enzyme* could be genetically inherited as most metabolic inborn errors, e.g., hypophosphatasia with low serum ALP level and Wilson's disease with low serum ceruloplasmin level. It could be acquired as low serum pseudocholinesterase level in hepatitis, and, low serum amylase level in chronic hepatic and pancreatic diseases and severe malnutrition.
- *Increased enzyme inhibition,* e.g., insecticide poisoning that leads to low serum pseudocholinesterase activity, but assaying the protein with immunoassays will show normal enzyme level.
- *Lack of cofactors,* e.g., pregnancy and liver cirrhosis displays low serum GOT level.

APPLIED EXAMPLES OF PLASMA ENZYME PATTERN (ENZYMOGRAM)

In heart diseases within the first day of infarction, elevation of serum CK is noticed followed by GOT and GPT (GOT is also called aspartate aminotransferase – AST, and, GPT is also called alanine aminotransferase - ALT) that peaks at 3rd days and LDH that peaks at 5th days. Other enzymes are also used and include; γ-glutamyl-transpeptidase (γGTP), histaminase, pseudocholinesterase and aldolase. However, the serum level of these enzymes also increases in non-cardiac diseases, e.g., CK in hypothyroidism, muscular dystrophy, and dermatomyositis; GOT in muscular and hepatic diseases; LDH in cancer, pulmonary embolism, renal diseases, pernicious anemia, and muscle and liver diseases; aldolase in dermatomyositis, muscular dystrophy and viral hepatitis, and, γGTP in hepatobiliary disorders, alcoholism and pancreatic diseases. The detection of tissue-specific isoenzyme would resolve confusion about tissue origin of some of these enzymes, e.g., cardiac MB-CPK and LDH-1 and -2.

In liver disease abnormally elevated levels of the following enzymes are detected; GPT, GOT, ALP (particularly in post-hepatic jaundice, cancer liver and metastatic carcinoma), 5'-nucleotidase (particularly in biliary tract diseases), LDH, isocitrate dehydrogenase (particularly in infective hepatitis, malignancy and drug toxicity), ornithine carbamoyl transferase (particularly in viral hepatitis, obstructive jaundice, cirrhosis and metastatic carcinoma) and sorbitol dehydrogenase (particularly in viral hepatitis and chemical poisoning). However, ALP is also elevated in rickets, osteomalacia, hyperparathyroidism, Paget's disease and bone malignancy

In gastrointestinal diseases, elevated serum pancreatic amylase and lipase levels are detected in acute pancreatitis, mumps, perforated peptic ulcer and intestinal obstruction.

In malignancy elevated serum levels of LDH, aldolase, phosphohexose isomerase are detected in widespread malignancies and leukemia; cathepsins, plasmin (serine endopeptidase; EC 3.4.21.7) and other proteases in metastatic tumors; LAP in liver carcinoma; acid phosphatase in prostate carcinoma, osteolytic metastasis from breast, leukemia and myeloproliferative disorders (but also in Gaucher's disease, hemolytic anemia, thrombocytosis, Paget's disease and pulmonary embolism); β-glucuronidase in cancer of urinary bladder, cancer head of pancreas, breast and cervix cancer; and alkaline phosphatase in liver and bone metastasis and carcinoma of pancreas.

Chapter XI

ENZYME ENGINEERING AND INDUSTRIAL APPLICATIONS OF ENZYMES

INTRODUCTION

Modern enzyme biotechnology began in 1874 when Christian Hansen extracting dried calves' stomachs with saline solution to prepare rennet for cheese manufacturing. However, enzymes were used for long time either in the form of vegetables rich in enzymes, or in the form of microorganisms, e.g., for brewing processes, in baking, and in the production of alcohol. Therefore, enzymes are the engineers of biotechnology and are engineered by biotechnology, e.g., amplifying their genes for larger production, and, by mutating their genes to be constitutively active, not to be inhibited by feedback effectors, to have high stability against temperature and pH changes, to withstand the organic environment, and, change substrate specificity.

Enzyme engineering (or enzyme biotechnology) is the usage of the catalytic activity of isolated enzymes, to produce new metabolites or to convert some compounds into another's (biotransformation) useful as chemicals, pharmaceuticals, fuel, food or agricultural additives. Natural source of these enzymes are animal tissues, plants, fungi and bacteria, and, cloned genes for specific recombinant enzymes production in foreign host organisms. Enzyme reactor consists of a vessel containing a reaction medium, used to perform a desired conversion by natural or recombinant enzymes. Enzymes used in this process are free in the solution or immobilized in particulate, membranous or fibrous support. However, because enzymes are limited in the number of reactions they have evolved to catalyze and because they lack stability in organic solvents and at high temperatures, new enzyme are engineered (through; i) *rational design*, or, ii) *molecular evolution*) to suite the requirements; see later.

EXTRA ATTENTION: THE GENERAL TYPES OF CATALYSIS

There are two general types of catalysis; i) *the homogeneous catalysis* where reactants and catalyst are in the same phase as most body reactions with catalysts and reactants occurring free in the aqueous environment, and, ii) *the heterogeneous catalysis* where the catalyst is in a different phase than the reactants and products, e.g., sold phase catalysis with catalyst fixed as a sold phase and reactants and products (in liquid or gas forms) are free to access or leave the catalyst at the solid-liquid interface.

Amylase - instead of the conventional acid hydrolysis - breaks starch into simpler sugars useful, e.g., for baking and high-fructose corn syrup preparation after isomerizing glucose into fructose using glucose isomerase. These syrups have enhanced sweetening properties and lower calorific values than sucrose for the same level of sweetness. Liquefaction of starch was also improved by using a heat-stable α-amylase. Likewise, proteases are used to lower flour protein content for a smoother biscuit manufacturing and to predigest baby food. Natural (from calf stomach) or recombinant renin is used to hydrolyze proteins during cheese manufacturing. Lipases are used during the production of Roquefort cheese to enhance the ripening of the blue-mould cheese, and, lactase hydrolyzes milk lactose for the usage of lactase-deficient people. Papain (a di and tri-peptidase from the papaya fruit; EC 3.4.22.2) is used to soften meat for cooking. Cellulase is used to break down cellulose into sugars that can be fermented into ethanol in biofuel production. As biological detergents and contact lens cleaners, proteases (e.g., the peptidase subtilisin; EC 3.4.21.62), amylase, lipase and cellulase are used. In rubber industry, catalase is used to generate O_2 from peroxide to convert latex into foam rubber. In photographic industry, a protease is used to dissolve gelatin off scrap film to recover its silver content. Textile desizing (starch removal) by amylase was used instead of the long difficult and textile damaging methods of treatment with acid, alkali or oxidizing agents, or soaked in water for several days so that naturally occurring microorganisms could break down the starch. Enzymes are also used for industrial and pharmaceutical production, e.g., synthesis of amino acids, nucleosides and nucleotides, antibiotics, steroids, etc.

In molecular biology reagents industry, restriction enzymes, nucleotide transferases, DNA ligase and polymerases are used to manipulate DNA in genetic engineering and polymerase chain reactions, important in pharmacology, agriculture, medicine (e.g., urokinase to activate intravascular blood clot dissolution through activating plasminogen into plasmin, and, encapsulated digestive pancreatic proteolytic enzymes in cases of pancreatic insufficiency, e.g., cystic fibrosis) and forensic science. Several enzymes are applied as reagents in laboratory diagnostic techniques, e.g., glucose, glycerol and cholesterol oxidases in determination of glucose, triglycerides or cholesterol levels in clinical samples. However, the utmost important application is the investment of enzyme kinetics and mechanisms in developing enzyme inhibitors as drugs that targets specific metabolic pathways as: antibacterial, antiviral, anticancer, and, antimetabolic drugs. Although designed to be specific for these conditions, the close similarity between metabolic enzymes in viruses, bacteria and cancer cells to the normal cells made it inevitable that the patient would succumb some side-effects.

EXTRA ATTENTION: ENZYME IMMOBILIZATION

Although all enzymes are synthesized inside cells, they are active inside and outside the cells *in vivo* and *in vitro* along the reaction conditions are optimized. For research investigations and/or industrial purposes, enzymes may be immobilized while active in three ways. Immobilization of enzymes by adsorbed or attachment to an inert insoluble material increases resistance of the enzyme to changes in conditions such as pH or temperature. This also allows easy recycling of the enzyme by withdrawal of the pure product without compromising the catalytic activity. This is applied for large scale catalysis by, e.g., acylases, lipases, proteases, invertase, etc.

The Immobilization is carried out by three methods:

- *Adsorption on the outside of an inert material*, e.g., glass, charcoal or alginate as gel, beads or matrix. It is much cheaper, simpler and commonly used, e.g., in zymography techniques, but it reduces the enzyme catalytic ability by reducing the accessibility of the active site of the immobilized enzyme.

- *Entrapment into insoluble beads or microspheres*, such as calcium alginate beads. This may hinders the availability of the substrate, and the exit of products.
- *Covalent cross-linkage to a matrix activated by a chemical reaction* that avoids the active site of the enzyme. This method is by far the most effective method. However the inflexibility of the covalent bonds precludes the self-healing properties exhibited by chemoadsorbed self-assembled monolayers. Use of a spacer molecule like polyethylene glycol reduces the steric hindrance by the substrate in this case.

Biosensors are composed of immobilized enzyme(s) that react with substrate to generate a product which is used by a transducer to generate an electrical signal. They express the rate of substrate consumption and/or product generation. An example is the glucose electrode that is composed of a layer of glucose oxidase immobilized on polyacrylamide gel around a platinum oxygen electrode. Contact with a solution containing glucose activates the reaction with O_2 to generate H_2O_2 and gluconolactone. The electrode detects the corresponding reduction in O_2. Different types of biosensors are used as detecting devices in diagnostic clinical biochemistry, food hygiene and detection of environmental pollution.

ENZYME ENGINEERING AND DESIGN

Generating enzymes with higher stability towards temperature, oxidation, organic solvents and harsh reaction environment and with higher catalytic abilities towards a selected substrate for mass biosynthetic and/or degradative applications utilizes the genetic-manipulation techniques for large-scale supply of such enzymes. They are tailor-made biocatalysts created from wild-type enzymes by mutation induced protein engineering that requires constructing the enzyme encoding gene, a suitable expression system (usually microbial), and a sensitive enzyme detection and characterization system. The process uses either of two approaches;

- Computer-aided rational molecular modeling design with site-directed mutagenesis.

- Directed molecular evolution techniques.

In the first approach, *the rational molecular modeling design of biocatalysts,* the process is information-intensive since it requires knowledge of the structure and the relationships between sequence, structure and mechanism/function so as to be able to increase the selectivity, activity and the stability of enzymes. The approach modifies the critical amino acid residues based on the understanding of the three-dimensional structure, i.e., those close to the active site and the binding pocket. It is composed of the following steps;

1. Understanding protein structure.
2. Rational site-directed mutagenesis based mainly on the understanding of the protein X-ray crystallography data..
3. Recombinant vector design and transformation of the host.
4. Protein expression and purification.
5. Protein characterization, e.g., for gene and protein sequencing, protein stability, kinetics, substrate binding and range of its specificity.
6. Selection of improved variants.

Rational protein design is useful for the reinforcement of a weak reaction, change of enzyme mechanism, substrate specificity, cofactor specificity, enantioselectivity, and stability, as well as the elucidation of enzyme mechanisms. Factors influencing thermostability have recently been elucidated by a structural comparison of various enzymes from mesophilic and thermophilic organisms. Examples of successful strategies to enhance thermostability are the removal of asparagine residues in α-amylase, the introduction of more rigid structural elements such as proline into α-amylase and D-xylose isomerase, or disulfide bridges to stabilize chicken egg lysozyme. The introduction of additional hydrophobic binding pocket contacts was shown to stabilize bacterial 3-isopropylmalate dehydrogenase and formate dehydrogenase.

To increase stability towards oxidation, removal of cysteine and methionine residues exhibited positive effects in the case of formate dehydrogenase, (R)-3-hydroxybutyrate dehydrogenase, and D-amino acid oxidase. However, examples of improved and inverted enantioselectivity of enzymes by this approach are rare compared to directed molecular evolution. In a bacterial pyruvate decarboxylase, mutation of Trp392, a bulky residue in the substrate-binding channel, increased the carboligase side-reaction of the enzyme by a factor of six, without negatively influencing the stability and the enantioselectivity. Removal of the sterically hindering carboxy-terminal tetrapeptide converted the aminopeptidase into an oligopeptidase. The substrate range of P450cam was extended from camphor to polycyclic aromatic hydrocarbons by mutation of two aromatic residues (Phe87 and Tyr96) in the substrate access channel. Alteration of substrate specificity was achieved by site-directed mutagenesis of human leukocyte 5-lipoxygenase, yielding a 15-lipoxygenating biocatalyst. A dicarboxylic amino acid β-lyase, a novel enzyme not found in nature, was generated by site-directed mutagenesis of a tyrosine phenol-lyase.

In the second approach, *directed molecular evolution* (evolutive biotechnology), either the random mutagenesis of the gene encoding the catalyst (e.g., by error-prone PCR), or recombination of gene fragments (e.g., derived from DNase degradation, the staggered extension process or random priming recombination) is used. The improved enzyme variants are selected from the created gene libraries. The approach creates variant gene sequences to get variant protein structures and then analyze them for altered activity. This very often revealed that mutations afflicting amino acids far away from the active sites (*irrational mutations*) are also fundamental to alter, e.g., substrate specificity of hydrolases). It is composed of the following steps;

1. Thousands of random mutagenesis (using error-prone PCR ad DNA shuffling).
2. Recombinant vectors carrying libraries of mutant genes and transformation of the host - particularly high mutator strains.
3. Protein expression and micro-characterization (e.g., in a microtiter plate format of the cultured bacteria) for stability, kinetics, substrate binding and range of its specificity.
4. Selection of improved mutants for protein purification and characterization, e.g., analysis of the nature of mutations acquired. *In vitro* recombination by DNA shuffling of selected mutants in a second round of directed molecular evolution could further improve them.
5. Mass expansion and protein production.

Directed evolution make the use of the ability of natural selection to evolve proteins or RNA with desirable properties not selected for in the natural organism. Directed evolution is performed *in vivo* through cloning into living cells to select for properties in a cellular context, or, *in vitro* for more versatile and larger scale of selections. Prior understanding of the mechanism of the useful activity is not a prerequisite in order to improve it for agricultural, medical or industrial applications.

Both protein engineering approaches can be repeated or combined until biocatalysts with desired properties are generated. An impressive example of the use of directed evolution and rational protein design targeted a heme peroxidase from a mushroom fungus to be used as a dye transfer inhibitor in laundry detergent. Variants produced by error-prone PCR and rational design were screening for improved stability by measuring residual activity after incubation under conditions mimicking those in a washing machine (e.g. pH 10.5, 50°C, 5 - 10 mM peroxide). Surprisingly, for both methods sequencing of the best variants identified position Glu239 to be crucial for success. Furthermore, replacement with glycine - as predicted by computer-modeling - gave the best performance. Subsequent *in vivo* shuffling led to dramatic improvements, yielding a mutant with 174 times the thermal stability and 100 times the oxidative stability of the wild-type peroxidase.

Phosphoribosylanthranilate isomerase activity was evolved from the α/β-barrel scaffold of indole-3-glycerol phosphate. The enantioselectivity of a bacterial lipase towards 2-methyldecanoate was increased in a mutant bearing five amino acid substitutions. The solved structure of this lipase showed that the increased enantioselectivity is caused by increasing the flexibility of distinct loops of the enzyme; however, none of the mutations are located near the binding pocket. A triple mutant of cytochrome P450 BM-3 obtained by directed evolution was found to hydroxylate indole, producing indigo and indirubin. The inversion of enantioselectivity of a hydantoinase, from D-selectivity to moderate L-preference by a combination of error-prone PCR and saturation mutagenesis required only one amino acid substitution. Thus, production of L-methionine from D,L-5-(2-methylthioethyl) hydantoin in a whole-cell system of recombinant E. coli that also contain an L-carbamoylase and a racemase, at high conversion became feasible. The substrate specificity of a peroxidase from Saccharomyces cerevisiae towards guaiacol was increased 300-fold by means of DNA shuffling.

It is often assumed that improving a biocatalyst in one direction affects other desired enzyme characteristics. It has been demonstrated, however, that it is possible to increase the thermostability of a cold-adapted protease to 60 °C (thermophilic) while maintaining high activity at 10 °C. The best psychrophilic (cold-hot adapted) subtilisin S41 variant contained only seven amino acid substitutions constituting only a tiny fraction of the usual 30–80% sequence difference found between psychrophilic enzymes and mesophilic (cold adapted) counterparts. Simultaneous screened for four properties - activity at 23 °C, thermostability, organic solvent tolerance and pH-profile - in a library of family-shuffled subtilisins revealed variants with considerably improved characteristics for all parameters.

Not only improving one specific biocatalyst, but also engineering of entire metabolic pathways by means of directed evolution is possible. The first example, phytoene desaturases and lycopene cyclases were shuffled in the context of a carotenoid biosynthetic pathway assembled from different bacterial species. Phospholipase activity was introduced into a Staphylococcus aureus lipase by directed evolution using error-prone PCR and gene shuffling. The best variant contained six mutations and displayed a 11.6-fold increase in phospholipase activity and a 11.5-fold increased phospholipase:lipase ratio compared to the wild type.

EXTRA ATTENTION: DNA SHUFFLING

DNA shuffling is a process in which DNA of a related gene family of enzymes is digested with restriction enzymes and the produce DNA fragments are subjected to PCR without adding primers. Instead, the single stranded DNA of unrelated parts of these genes would be complementary at random areas and bind in a staggered manner to leave non-complementary hanging-over sequences at both ends. These ends would be used as template for the polymerase reaction. The PCR product would be a chimeric DNA double strand from two unrelated gene fragments. The produce new chimeric fragments are ligated into new shuffled full-length genes.

Chapter XII

THE ENZYME AS DRUGS: PRIMARY AND REPLACEMENT THERAPIES

INTRODUCTION

Enzymes as therapy are either used to replenish a missing enzyme due to an inherited gene defect or as a primary therapy, i.e., unrelated to such diseases. Enzymes as drugs are specific to substrate and catalytically highly active on such substrate. A therapeutic enzyme was first described as part of replacement therapies for genetic deficiencies in the 1960s. They are administrated through injection, topical application, inhalation and orally. The first recombinant enzyme as a drug, was the clot dissolving Activase 1 (alteplase) - the recombinant human tissue plasminogen activator, was approved for human heart attacks in 1987. Polyethylene glycol-conjugated bovine adenosine deaminase (Adagenl) was approved in 1990, to treat patients afflicted with inherited adenosine deaminase deficiency type of severe combined immunodeficiency disease (SCID). Because of their specificity and potency, therapeutic enzymes now cover a wide range of diseases and conditions that include; inherited diseases (e.g., Gaucher's, Fabry's, mucopolysaccharidoses I, II and VI, Pompe's glycogen storage, cystic fibrosis, phenylketonuria, and adenosine deaminase deficiency), pro- and anticoagulants, antineoplastic enzymes and prodrug activator enzymes, antifungal, antiprotozoal and antibacterial enzymes, burn debridement and others. Pompe's disease was the first muscle disorder to be treated by enzyme replacement therapy.

The replaced adenosine deaminase conjugated to polyethylene glycol has enhanced half-life (originally less than 30 min) and reduced possibility of immunological reactions due to the bovine origin of the drug. The enzyme cleaves the excess circulating adenosine of the patients to reduce its toxicity to the immune system. Ceredasel (alglucerase injection; glucocerebrosidase) for the treatment of Gaucher's disease, a lysosomal storage disease, was the first enzyme replacement therapy in which an exogenous enzyme was targeted to its correct compartment within the body. First the source was a modified placental glucocerebrosidase (Ceredasel) and subsequently recombinant human enzyme (imiglucerase) was used. The second lysosomal storage disease to follow was Fabry's disease; a fat (glycolipid) storage disorder caused by a deficiency in α-galactosidase. It primarily affects the vasculature and results in renal failure, pain, and corneal clouding. Recombinant human α-galactosidase was used. Chondroitinases promote regeneration of injured spinal cord by removing, in the glial scar, the accumulated chondroitin sulfate that inhibits axon growth. Hyaluronidase has a similar hydrolytic activity on chondroitin sulfate and may also help in the regeneration of damaged nerve tissue.

The holistic oral digestive enzyme extracts were long in use. Congenital sucrase-isomaltase deficiency, for example, is treatable with sacrosidase - a β-fructofuranoside fructohydrolase from Saccharomyces cerevisiae that can be taken orally. Phenylketonuria (PKU) is another genetic disorder requiring strict compliance with a specialized diet. PKU is caused by low or non-existent phenylalanine hydroxylase activity, which catalyzes the conversion of phenylalanine to tyrosine. An oral treatment, Phenylase™, is a recombinant yeast phenylalanine ammonia lyase that is able to degrade phenylalanine in the gastrointestinal tract. A mixture of pancreatic enzymes, including lipases, proteases and amylases, has been shown to be useful in the treatment of fat malabsorption in patients with human immunodeficiency virus and pancreatic insufficiency in cystic fibrosis patients.

Other possible enzyme treatments for other digestive diseases include oral peptidase supplement therapy could be used for the treatment of Celiac Sprue (Celiac disease), a widely prevalent disorder of the small intestine caused by an immune reaction to the gliadin protein in ingested wheat products. Inhalable enzyme formulations were applied to cystic fibrosis. Pulmozymel (Dornase a), a DNase, liquefies accumulated mucus in the lung and diminishes pulmonary tissue destruction by lowering the level of matrix metalloproteinases in the bronchoalveolar lavage fluid.

Lysozyme has been used as a naturally occurring antibacterial agent in many foods and consumer products, because of its ability to break carbohydrate chains in the cell wall of bacteria. Lysozyme has also been shown to possess activity against HIV, as has RNase A and urinary RNase U, which selectively degrade viral RNA. Other naturally occurring antimicrobial agents are the chitinases. As an element of the cell wall of various pathogenic organisms, including fungi, protozoa and helminthes, chitin is a good target for antimicrobial.

Polyethylene glycol-conjugated arginine deaminase, an arginine-degrading enzyme, can inhibit human melanoma and hepatocellular carcinomas, which are auxotrophic for arginine owing to a lack of arginosuccinate synthetase activity. Polyethylene glycol-conjugated asparaginase, Oncaspar (pegaspargase), is used in the treatment of children with newly diagnosed acute lymphoblastic leukemia similar to the asparagine-degrading enzyme bacterial asparaginase - since these cancer cells are auxotrophic for asparagine. They are effective adjuncts to standard chemotherapy. The removal of chondroitin sulfate proteoglycans by chondroitinase AC and, to a lesser extent, by chondroitinase B, inhibits tumor growth, neovascularization and metastasis. Antibody-directed enzyme prodrug therapy in which a monoclonal antibody carries an enzyme or the enzyme itself has antibody-like targeting domains targets and kills specifically cancer cells - where the enzyme activates a prodrug to specifically destroying cancer cells but not normal cells. One of the side-effects of cancer chemotherapy is hyperuricemia, a build-up of uric acid that results in gouty arthritis and chronic renal disease. Urate oxidase is able to degrade the poorly soluble uric acid. Interestingly, the gene for this enzyme is present in humans, but possesses a nonsense codon. Recombinant Rasburicase (Eletik) is safe and effective as uricolytic agent particularly in its polyethylene glycol conjugated. Future investigations concerning antiageing, organ injury in hemorrhagic shock, stoke and ischemia-reperfusion injuries concentrate on usage of the antioxidant superoxide dismutase and catalase enzymes since they prolonged the life of Caenorhabditis elegans. Human butyryl-cholinesterase, a naturally occurring serum detoxification enzyme, acts to break down acetylcholine and could be useful for the treatment of cocaine overdose. Structure-based engineering and directed evolution of the enzyme has resulted in higher activity toward cocaine.

CONCLUSION

The book highlighted the outmost importance of understanding enzymology and its applications particularly in medicine for undergraduate and graduate medical, pharmacy, dentistry, biotechnology and biology students. It covers all the fundamental aspects of the science in a very simpilifies manner with rational building up of information. In every level it relates information to applied examples and investment. This book is hoped to establish a proper understanding for enzymology within the medical and biology students' anena that is particularly important in this era of medical, pharmaceutical and biotechnological applications of enzymes.

Review Questions

Review Questions

I. Write Briefly on:

1. EC classification of enzymes.
2. List seven coenzymes derived from vitamins.
3. Role of metals in enzymes.
4. Metalloenzymes and Metal-activated enzymes (with examples).
5. Allosteric enzymes.
6. Binding sites.
7. Multifunction enzymes and multienzyme complexes (with examples).
8. Substrate channeling.
9. Specificity and specificity constant.
10. Catalytic triad of serine proteases.
11. Significance of Km.
12. Immobilized enzymes.
13. Suicide inhibitors.
14. Industrial applications of enzymes.
15. Mechanism-based inactivators and drug design.

16. Abzymes.
17. Organophosphorus poisoning and enzymes.
18. Drugs as competitive inhibitors of enzymes.
19. Feedback inhibition.
20. Inducible enzymes.
21. Coupled enzyme assays.
22. Enzyme engineering and design.
23. Enzymes replacement therapy.

II. Discuss the Following:

1. Co-enzyme can be considered as co-substrate.
2. Induced fit model is the most and accepted enzyme-substrate-coenzyme binding model.
3. Allosteric enzymes display sigmoidal substrate-activity curves.
4. Some enzymes operate with kinetics faster than diffusion rates.
5. The requirement of activation energy is essential in biological systems.
6. Some enzymes show shift in their substrate specificity.
7. Monooxygenases are known as 'mixed function oxidases'.
8. Drugs can be designed based on knowledge of substrate binding and reaction mechanisms.
9. Sigmoid kinetic reflects cooperative interactions between protein subunits mediated by non-covalent bonds.
10. Rapid disposal of end products is essential for the continuation of the metabolic pathways.
11. Functional and non-functional plasma enzymes.
12. Sulfonamides act as antibiotics.

13. Non-competitive inhibition is a special type of mixed inhibition.
14. Concerted and Cumulative feedback inhibition.
15. Some enzymatic covalent modification is irreversible.
16. A number of enzymes are secreted as zymogens.
17. Isoenzymes differ in their physical characteristics.
18. Enzyme engineering gives rise to new characteristics and/or new activity to the enzyme.

III. Multiple Choice Questions

1. In the reaction; β-Carotene + $O_2 \Rightarrow$ Retinal; the catalyzing enzyme is:
   A. Dihydroxylase.
   B. Dioxygenase.
   C. Dioxidase.
   D. All of the above.

2. In the reaction; 2 GSH + $H_2O_2 \Rightarrow$ GSSG + 2 H_2O; the catalyzing enzyme is:
   A. Oxidase.
   B. Catalase.
   C. Peroxidase.
   D. None of the above.

3. A conformation of an enzyme is:
   A. Flexible structure.
   B. A defined shape and volume.
   C. A defined volume.
   D. All of the above.

4. If equilibrium constant of a reaction is very high:
   A. Energy is consumed by the reaction.
   B. Energy of reaction doesn't change.
   C. Energy is released by reaction.
   D. All the above.

5. A graph of activity versus ---------- is bell shaped:
   A. Activator.

   B. Inhibitor.
   C. pH
   D. All the above.

6. According to Michaelis-Menten, Km of an enzyme is:
   A. Substrate concentration at maximal rate.
   B. Substrate concentration at half maximal rate.
   C. Can only be determined after by linearization.
   D. All the above.

7. Which one of the following can be a measure of enzyme specificity:
   E. Km.
   F. Kcat.
   G. Kcat/Km.
   H. Km/Kcat.

8. Mechanism of an enzymatic reaction can be deduced from:
   A. X-Ray structure of enzymes.
   B. Kinetic studies.
   C. Effect of pH.
   D.
   E. All of the above.

9. A Coenzyme that is not a vitamin is:
   A. S-Adenosyl methionine.
   B. Coenzyme-Q.
   C. Lipoic Acid.
   D. All the above.

10. Enzymes enhance the rate of the reactions by all of the following, EXCEPT:
    A. Proximity and orientation.
    B. Enhancing the equilibrium constant.
    C. By providing residues for acid-base catalysis.
    D. Covalent catalysis.

11. An inhibitor that increases Km of an enzyme is:
    A. Non- competitive inhibitor.
    B. Competitive inhibitor.
    C. Uncompetitive inhibitor.
    D. Mixed inhibitor.

12. The characteristics of Competitive inhibitors include:

A. They are structurally analogous of substrates.
B. They have no effect on V_{max}.
C. They are mutually exclusive with substrate.
D. ll the above.

13. A non competitive inhibitor of cyclooxygenase in prostaglandin synthase is:
    A. Aspirin.
    B. Arachidonic acid.
    C. Amino acid.
    D. Lipoic Acid.

14. Cyanide is which type of inhibitor:
    A. Coenzyme inhibitor.
    B. Inhibitor of specific ion cofactor.
    C. Prosthetic group inhibitor.
    D. Apoenzyme inhibitor.

15. Allopurinol is a competitive inhibitor for:
    A. Purine Synthetase.
    B. Xanthine oxidase.
    C. Cyclooxygenase.
    D. Glycopeptide transpeptidase.

16. Heavy metal toxicity is caused by:
    A. Replacement of the metal ions at the active site.
    B. Denaturation of the enzyme by the metal ion.
    C. Dissociation of the prosthetic group from the enzyme.
    D. Binding with the functional groups at the active site.

17. All of the following are non-competitive inhibitors, EXCEPT:
    A. Aspirin for cyclooxygenase.
    B. Malonic acid for Succinyl dehydrogenase.
    C. AMP for fructose-1,6-diphosphatase.
    D. Cyanide for cytochrome oxidase.

18. Which of the following is not true for allosteric regulators:
    A. They could be positive or negative modulators.
    B. Little or no similarity to the substrates or coenzyme.
    C. They alter the conformation of the enzyme.
    D. They always decrease the Vmax of the enzyme.

19. According to the concerted model of enzyme regulation:

A. The enzyme follows 'All or None' pattern.
B. The conformation change is sequentially affected.
C. The modulator binds with the subunits with increasing affinity.
D. The enzyme can exist in multiple conformation states.

20. Which of the following is NOT true about covalent modification of enzymes:
    A. It causes an activating/inactivating conformational change.
    B. Phosphorylation can increase or decrease the activity of enzymes.
    C. Covalent modification results in increased synthesis of the enzyme.
    D. Covalent modification of enzymes is mostly reversible.

21. Gene mutation that severely affected enzyme ability to bind a coenzyme; as a consequence:
    A. The enzyme will not bind the substrate.
    B. The formation of the transition state complex will be prevented.
    C. An alternative coenzyme will be used.
    D. The disease is ameliorated by increasing dietary coenzyme vitamin precursor.

Answers key for MCQs:

1, B; 2, C; 3, A; 4, C; 5, D; 6, B; 7, C; 8, D; 9, A; 10, B; 11 B; 12, D; 13, A; 14, C; 15, B; 16, D; 17, B; 18, D; 19, A; 20, C; 21 B.

REFERENCES AND FURTHER READING AND WEB-BASED RESOURCES

REFERENCES AND FURTHER READING RESOURCES:

[1] Illanes A. (editor): *Enzyme Biocatalysis: Principles and Applications*; 2008, Springer Science.

[2] Barrett A, Rawlings N, Woessner J (editors). *Handbook of Proteolytic Enzymes*, 2nd ed., 2004. Academic Press, NY.

[3] Marangoni AG (editor). *Enzyme Kinetics: A Modern Approach*, 2002. John Wiley & Sons, Weinheim.

[4] Suckling CJ, Gibson C, and Pitt A (editors). *Enzyme Chemistry: Impact and Applications*, 3rd ed, 1998. Kluwer Academic Publishers, Amsterdam.

[5] Bisswanger H (editor). *Enzyme Kinetics; Principles and Methods*. 2nd Ed., 2008 WILEY-VCH Verlag GmbH & Co. KGaA, Weinheim.

[6] Smith H and Simons C (editors). *Enzymes and Their Inhibition: Drug Development*, 2005. CRC Press, NY.

[7] Reymond J (editor). *Enzyme Assays: High-throughput Screening, Genetic Selection and Fingerprinting*, 2006. John Wiley & Sons, Weinheim.

[8] Buchholz K, Kasche V and Bornscheuer UT (editors). *Biocatalysts and Enzyme Technology*, 2005. John Wiley-VCH, NY.

[9] Drauz K and Waldmann H (editors): *Enzyme Catalysis in Organic Synthesis*; 2002, 2nd Ed., Wiley-VCH Verlag GmbH, Weinheim.

[10] Ptashne H and Gann A (editors). *Genes and Signals: Imposing Specificity on Enzymes by Recruitment,* 2002. Cold Spring Harbor Laboratory Press, NY.

[11] Breslow R (editor). *Artificial Enzymes*, 1st ed., 2005. Wiley-VCH, NY.

[12] Copeland RA (editor). *Enzymes: A Practical Introduction to Structure, Mechanism, and Data Analysis*, 2nd ed., 2000. Wiley-VCH, Weinheim.

[13] Copeland RA (Editor): *Enzymes: A Practical Introduction to Structure, Mechanism, and Data Analysis*; 2nd ed., 2000, Wiley-VCH, Inc., NY.

[14] Brakmann S and Johnsson K (editors). *Directed Molecular Evolution of Proteins, or How to Improve Enzymes for Biocatalysis*, 2002. Wiley-VCH, NY.

[15] Deshpande S (editor). *Enzyme Immunoassays: From Concept to Product Development*, 1996. Kluwer Academic, Amsterdam.

[16] Bornscheuer UT and Pohl M. *Improved biocatalysts by directed evolution and rational protein design.* Current Opinion in Chemical Biology 2001, 5:137–143.

[17] Rodwell VW and Kennelly PJ (2003): *Enzymes: Mechanism of Action, kinetics; regulation of activity.* Section I (7, 8, and 9). 49-79. In; Harper's *Illustrated Biochemistry*, 26th ed., 2003 (Murray RK, et al., editors). Lange Medical Books/McGraw-Hill (Medical Publishing Division), New York.

Relevant web-based Resources:

[1] http://path.upmc.edu/cases/index.html

[2] http://www.qub.ac.uk/cm/cb/text/studgide

[3] http://www.labtestsonline.org

[4] http://www.indstate.edu/thcme/mwking/enzyme-kinetics.html

[5] http://www-biol.paisley.ac.uk/kinetics/contents.html

[6] http://www.ifcc.org/ifcc.asp

[7] http://www.ebi.ac.uk/thornton-srv/databases/enzymes/

[8] http://us.expasy.org/enzyme/

[9] http://www.emc.maricopa.edu/faculty/farabee/BIOBK/BioBookEnzym.html

[10] http://en.wikipedia.org/wiki/MetaCyc

[11] http://academicearth.org/lectures/enzyme-structure-and-function-1.

INDEX

B

C

D

E

F

G

R

S

T

U

V

W

X

Y

Z